Statistics

Wiley Self-Teaching Guides teach practical skills from accounting to astronomy, management to mathematics. Look for them at your local bookstore.

Other Math and Science Wiley Self-Teaching Guides

Math

All the Math You'll Ever Need, by Steven Slavin

Business Statistics, by Donald J. Koosis

Geometry and Trigonometry for Calculus, by Peter H. Selby

Practical Algebra, by Peter H. Selby and Steve Slavin

Quick Algebra Review, by Peter H. Selby and Steve Slavin

Quick Arithmetic, by Robert A. Carman and Marilyn J. Carman

Quick Calculus, by Daniel Kleppner and Norman Ramsey

Science

Astronomy, by Dinah Moché

Basic Physics, by Karl F. Kuhn

Biology, by Steven D. Garber

Chemistry: Concepts and Problems, by Clifford C. Houk and Richard Post

Statistics

A Self-Teaching Guide

Fourth Edition

Donald J. Koosis

John Wiley & Sons, Inc.

New York • Chichester • Weinheim • Brisbane • Singapore • Toronto

To Irene with even more love.

Library of Congress Cataloging-in-Publication Data

Koosis, Donald J.
    Statistics : a self-teaching guide / Donald J. Koosis. — 4th ed.
        p.   cm.
    Includes index
    ISBN 0-471-14688-9 (pbk. : alk. paper)
    1. Statistics—Programmed instruction. I. Title
QA276.12.K667 1997
519.5′07′7—dc21   96-50887

Printed in the United States of America
14 13 12 11

Contents

Preface

This book is designed for the individual who wants to learn the basic concepts and procedures of statistics as it is used today for educational, professional, or personal reasons. You may be interested in statistics as part of a liberal education. You may be a student of or employed in the field of education, engineering, or one of the social, biological, or physical sciences. The methods taught here are common to all these fields. You will find examples of the application of statistical techniques to each of these fields in the book.

This is a "how to" book. It will teach you how to perform a number of useful statistical tests and procedures. It will teach you how to determine which of these tests and procedures are appropriate under given circumstances. Although you will learn the basic concepts on which the tests are based, you will not learn the mathematical derivation of every formula nor all the possible variations of every test. The book assumes a knowledge of elementary algebra only—enough to substitute numbers for symbols in a formula and to keep the + and – signs straight when you are working with positive and negative numbers.

In this book, we emphasize the kinds of statistical work that can be done "by hand"—problems with small amounts of data and relatively simple computations. Such problems can be completed with a pencil and paper or a simple pocket calculator.

If you are reading this book, you probably have access to a personal computer or workstation that runs a spreadsheet program such as Microsoft Excel, Lotus 1-2-3, Corel QuattroPro, or Applix Spreadsheets. The formulas and procedures introduced in this book are easily implemented with a spreadsheet program. To make the transition even easier, the book includes explanations of the built-in spreadsheet functions that automate common statistical computations in two widely used spread-sheet programs—Microsoft Excel 5.0 and Lotus 1-2-3 4.0.

You will be able to use these techniques for yourself in everyday work and study, whether you use pencil and paper, a calculator, or a spreadsheet program.

As you work through this book, remember that the results of statistical tests and procedures are only as meaningful as the data on which they are based. The problems of measurement and experimental design are beyond the scope of this book and are touched on only in passing. In many cases you will have to apply knowledge of your field from other sources when you begin to put your statistical skills to use. At the end of the book is a test that you may use as a final examination to evaluate your progress. It includes material from all chapters. If you successfully complete the review problems for each chapter and the final test, you may be confident that you have mastered the material in this book.

How to Use This Book

This book is organized into numbered sections called frames. Each frame presents some new material, asks you to answer a question, and gives the correct answer. Most of the time you should be able to answer correctly. If you find that you have made an error, look back at the preceding material to make sure you understand the correct answer before you go on to the next frame.

This is not a book you can read through quickly. Statistics, like other mathematical subjects, requires concentration. Each chapter represents about two hours' work. If possible, try to complete a chapter in one or two sessions and stop only at the end of a section or chapter. Frequent interruptions make the material more difficult to master.

All the computations required in this book can be worked out by hand with reasonable effort; however, there is no special virtue in doing the arithmetic by hand. If possible, you should use a pocket calculator or a spreadsheet program.

If your interest is primarily in understanding the steps involved in a test, and you have only limited patience for arithmatic, you may choose to skip some of the computations. When you make this choice, be sure nevertheless that you set up the problem and compare your work, as far as it goes, with the solution in the answer.

At the end of each chapter is a set of review problems. These problems not only review the content of the chapter; they also relate it to other material you have learned. Do not skip them. Working these problems is an important part of the study process.

At the end of the book is a test that you may use as a final examination to evaluate your progress.

1   **Basic Skills**

This chapter is about how to summarize data. When you perform an experiment, conduct a survey, or collect information, you usually wind up with a number of observations. For example, you have the test scores of twenty-seven classmates, or the opinions of twelve jurors, or the height of one hundred redwood trees. Your first problem is to summarize this information in some way that lets you generalize—that lets you see the forest as well as the individual trees.

One common and useful way to summarize a group of observations is to draw a graph—a *frequency-distribution* graph. Another way is to compute some sort of "average" that describes a typical observation—a measure of central tendency. Three commonly used measures of central tendency are the *mean*, the *median*, and the *mode*.

It is also useful to describe numerically how much the observations differ from one another—what we call their *variability*. Three common measures of variability are the *range*, the *standard deviation*, and the *variance*.

This chapter will teach you the most common methods of summarizing data. When you have completed this chapter, you will be able to:

- Construct a frequency-distribution graph

- Recognize and apply some common vocabulary to describe frequency distributions

- Compute three measures of central tendency: the mean, the median, and the mode

- Compute three measures of variability: the range, the standard deviation, and the variance

- Do all of the above with a personal computer spreadsheet program

HOW TO CONSTRUCT A FREQUENCY-DISTRIBUTION GRAPH

The frequency distribution is a useful summary of most kinds of data. A frequency distribution sorts observations into categories and describes how often observations fall into each category (either as a number or a percent). Very likely you have used frequency distributions to analyze the data of your own daily life.

1 Consider the following data:

Irene has green eyes	Don has blue eyes
Tom has brown eyes	Carl has blue eyes
Marie has brown eyes	Sarah has brown eyes
Joan has blue eyes	Carol has brown eyes
David has brown eyes	Mark has brown eyes

How would you summarize these data?

Answer: You probably used a table like this:

Brown eyes	6, or 60%
Blue eyes	3, or 30%
Green eyes	1, or 10%

You might also have drawn a frequency-distribution bar graph like the one below. This type of graph is called a frequency-distribution *histogram*.

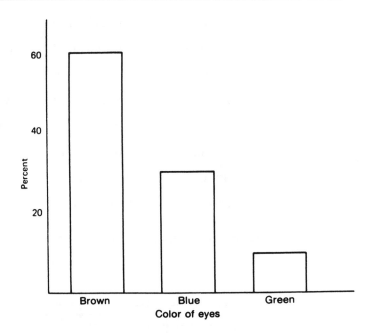

FIGURE 1-1:

Frequency distribution of color of eyes for 10 people

2 A bar graph like the one above is a useful summary of the data in question. This bar graph shows the _____ of eye colors.

Answer: Frequency distribution

3 Suppose you are doing market research for a company that sells automotive accessories. You are interested in the types of cars driven by customers of a particular mall. On a given afternoon, you count the following types of cars in the lot:

Family sedan	Economy car	Economy car
Utility/van	Family sedan	Wagon
Wagon	Family sedan	Utility/van
Sports car	Economy car	Utility/van
Family sedan	Family sedan	Pickup truck

Sports car	Utility/van	Wagon
Utility/van	Utility/van	Wagon
Sports car	Family sedan	Economy car
Economy car	Economy car	Wagon
Luxury sedan	Economy car	Utility/van
Classic car	Family sedan	Family sedan
Utility/van	Family sedan	Family sedan
Utility/van	Utility/van	

Prepare a frequency-distribution bar graph of these observations.

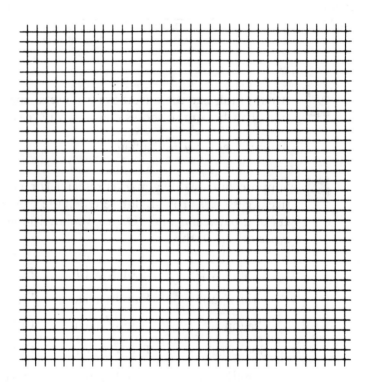

FIGURE 1-2:
Your graph paper

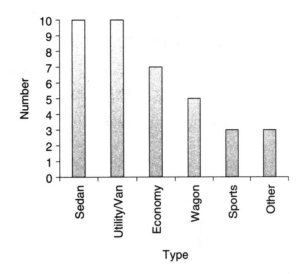

FIGURE 1-3:

Frequency distribution
of cars

Answer: Your graph should look something like the one above. As is customary, this graph measures the number of observations as a percentage along the vertical axis of the graph and shows the categories along the horizontal axis. The order in which you list the categories is not critical, although it helps the reader if there is some logic to the order; for example, here we have listed the categories from highest to lowest frequency.

4 Often, instead of categories, the data we wish to summarize are measurements on a continuous scale; for example, length or time or temperature. Then we must group the measurements into categories.

A common way of stating category boundaries is to make each category consist of observations *equal to or greater than* the lower boundary and *less than* the upper boundary. If you follow this convention, the labels on your graph are easier to read, but you and your readers must be clear about what the rule is for setting the boundaries.

For example, look at the following raw data and the frequency distribution based on these temperatures (in degrees):

59.0	62.0	70.0	72.5	74.5
59.5	62.0	70.5	72.5	74.5
60.5	63.5	71.0	73.0	74.5
61.0	63.5	71.0	73.0	75.0
61.0	64.0	71.5	73.0	75.0

61.0	67.5	72.0	73.0	75.0
61.0	68.0	72.0	73.5	76.0
62.0	69.5	72.0	73.5	76.5
62.0	70.0	72.5	73.5	77.0
62.0	70.0	72.5	74.0	79.0

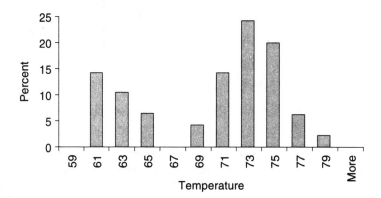

FIGURE 1-4:
Frequency distribution
of temperatures

The first bar in the graph above represents measurements greater than or equal to _____ and less than _____ degrees.

Answer: Greater than or equal to 60° and less than 62°

5 The second bar in the graph above represents measurements greater than or equal to _____ and less than _____ degrees.

Answer: Greater than or equal to 62° and less than 64°

6 In figure 1-4, how many categories are used to group the data?

Answer: 10

7 The following illustrations (graphs A, B, C) show three frequency-distribution graphs based on the same data. One is the original graph, which uses ten categories, another uses forty categories, and the third uses only three categories. Which one do you think presents the most useful summary of the data?

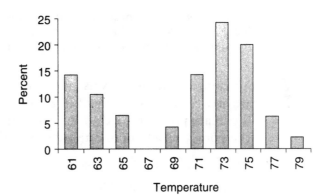

FIGURE 1-5:
Graph A

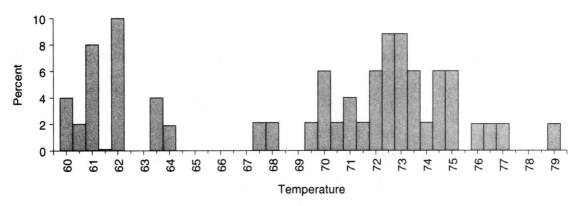

FIGURE 1-6:
Graph B

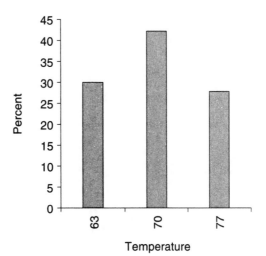

FIGURE 1-7:
Graph C

Answer: For most purposes graph A probably presents the most useful summary. Graph B is so detailed that the over-all pattern is not clear, and graph C goes to the other extreme and obscures a difference that is large enough to be important.

8 As a rule of thumb, it is usually best to choose categories so that there are between ten and twenty categories in a distribution of measurements. If you are preparing a distribution of test scores that range from 50 to 100 points, which set of categories would be preferred?

(a) 50–52, 53–54, 55–56, etc.

(b) 50–59, 60–69, 70–79, etc.

(c) 50–54, 55–59, 60–64, etc.

Answer:
(c) 50–54, 55–59, 60–64, etc.

9 By using ten categories, construct a frequency-distribution graph for the following data:

1.1	1.8	2.1	2.2	2.7	3.6
1.4	1.8	2.1	2.3	2.9	3.8
1.4	1.9	2.1	2.4	2.9	4.1
1.4	1.9	2.1	2.4	3.0	4.5
1.7	1.9	2.1	2.5	3.2	4.9

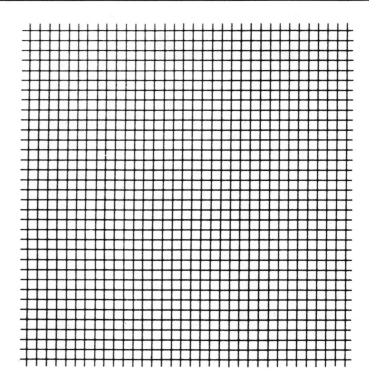

FIGURE 1-8:
Your graph paper

Answer: Your graph should look something like this one.

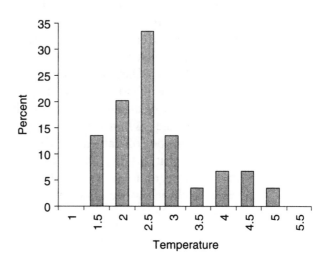

FIGURE 1-9:
Our frequency
distribution

FREQUENCY DISTRIBUTIONS WITH SPREADSHEET PROGRAMS

Most personal computer spreadsheet programs provide a way to generate a frequency distribution automatically from data in a worksheet. Refer to Figure 1-10 below, which could equally well represent Microsoft Excel or Lotus 1-2-3.

	A	B	C	D	E
	Data Range		Bin Range	Distribution Data	
2	0.5		1.0	2	
3	0.9		2.0	3	
4	1.0		3.0	1	
5	1.1			1	
6	1.2				
7	1.3				
8	2.1				
9	3.2				

FIGURE 1-10:
Frequency distribution in a spreadsheet

Not surprisingly, the first step in creating a frequency distribution is to enter the data in a column of the worksheet. This column of data is called the *data range*.

The next step is to define the categories into which the data will be classified. In spreadsheet jargon, these categories are called *bins*. To define bins, you enter the maximum values for each category in increasing order in another column of the worksheet, leaving a column of blank spaces to the right of the bin definitions.

Select Tools | Data Analysis | Histogram (for Microsoft Excel) or Range | Analyze | Distribution (for Lotus 1-2-3).

Complete the dialog to indicate the input data range and the bin definition range. When you select OK, the spreadsheet program will count the number of data items that fall into each bin and list this frequency distribution data in the column to the right of the bin descriptions. Any data that fall beyond the limit of the last bin definition will be listed in a cell below the data for the last bin.

10 The input data in figure 1-10 are in cells _____ . The bin definitions are in cells _____ . The information in cell D5 represents the number of observations greater than _____ .

Answer:

A2 through A9
C2 through C4
3.00

Once the data have been sorted, you may also create a chart or graph using the appropriate features of your spreadsheet software.

DESCRIBING FREQUENCY DISTRIBUTIONS

There are some technical terms commonly used to describe frequency distributions. You will find it useful to know these terms: *mode, bimodal, skewed,* and *normal.*

11 Distributions of measurements can have various shapes. Some of the more common situations are illustrated in figure 1-11. Assume that each of these distributions represents the times required by a group of 100 college students to solve a given Chinese puzzle. The time limit for each puzzle was half an hour.

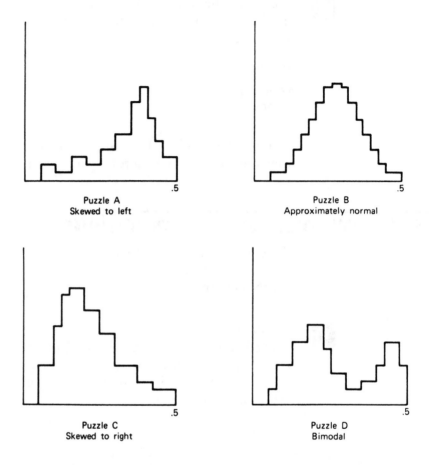

FIGURE 1-11:

Time required to solve
four puzzles, data
from 100 people

For one of these puzzles there were two obvious approaches to solving it. About half the students started with the correct approach and solved it quite quickly, whereas the other half attempted the wrong approach first and took substantially longer. Which puzzle meets this description and what name is applied to the distribution?

Answer:

Puzzle D
Bimodal

12   For one of the puzzles there was one clear average time. About as many finished faster than average as finished slower than average, and the distribution was bell shaped. Which puzzle fits this description and what name is applied to the distribution?

Answer:

Puzzle B
Approximately normal

> *The normal distribution has a precise mathematical definition and you will learn more about it later.*

13 Working with just one of the puzzles for a fixed amount of time automatically resulted in a solution. Most students simply kept working on the puzzle until it solved itself at the end of this time, but a moderate number were able to come up with faster solutions. Which puzzle fits this description and what name is applied to its distribution?

Answer:

Puzzle A
Skewed to the left

14 The peak, or mode, of a distribution is the most common score; that is, the tallest bar in the graph. What is the mode of this set of data?

3, 3, 4, 4, 4, 5, 5, 5, 5, 5, 5, 5, 7, 8, 8, 9, 9

Answer: The mode is 5, the most common score.

15 A peak in a distribution is called a mode. When a distribution has two peaks, as shown in figure 1-12, it is called _____ .

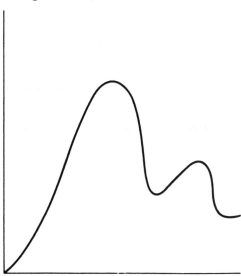

FIGURE 1-12:
A bimodal distribution

Answer: Bimodal

16 Look at the distributions below:

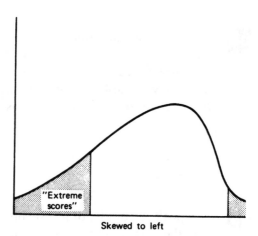

FIGURE 1-13:

Skewed to the left

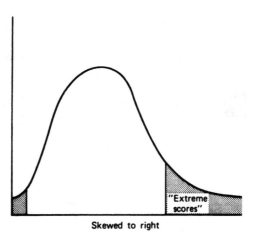

FIGURE 1-14:

Skewed to the right

When there is a greater number of extreme scores on the left side of a distribution than on the right side, we may say it is skewed to the (left/right).

Answer: Left

17   Sketch a distribution that is skewed to the right.

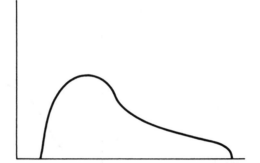

FIGURE 1-15:
Skewed to the right

18   Sketch an approximately normal distribution.

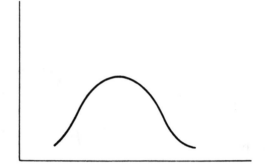

FIGURE 1-16:
Approximately normal

19   Sketch a bimodal distribution.

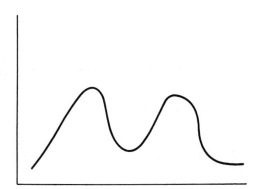

FIGURE 1-17:
Bimodal

20 What is the mode of this distribution?

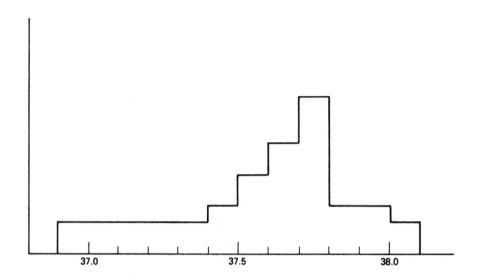

FIGURE 1-18:
What is the mode?

Answer: 37.7–37.8

21 What is the mode of this distribution?

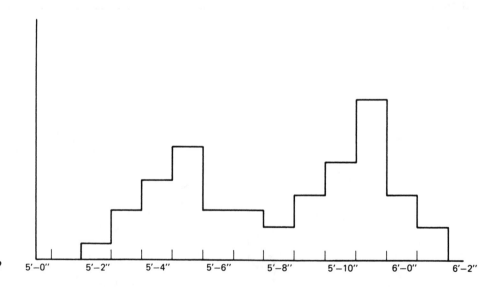

FIGURE 1-19:
What is the mode?

Answer: 5 ft. 5 in. and 5 ft. 11 in.

22 Describe each of these distributions in a few words.

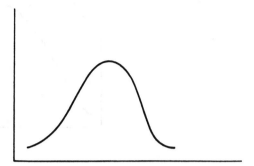

FIGURE 1-20:
Describe me

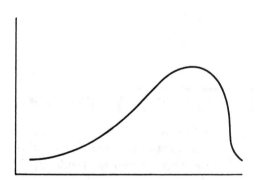

FIGURE 1-21:
Describe me

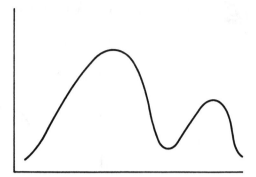

FIGURE 1-22:
Describe me

Answer:

(a) Approximately normal

(b) Skewed to the left

(c) Bimodal

MEASURES OF CENTRAL TENDENCY

Often we want to summarize a whole distribution of measurements with a single number that is typical of the distribution. This number is called a measure of central tendency. There are three measures of central tendency you should know—the mean, the median, and the mode. You already know about the mode, so you have only the median and the mean left to learn.

23 The *median* is the middle measurement in a set of measurements. To find the median you must

(a) Sort the observations in order of magnitude

(b) Then find the middle number

For example, in the following set of eleven measurements, the sixth is the median:

11, 11, 13, 15, 17, 17, 17, 19, 19, 19, 19

The median of this distribution is _____ .

Answer: 17

24 What is the median of the following set of measurements?

2.0, 5.8, 6.1, 10.5, 53.9, 54.0, 78.6

Answer: 10.5

25 What is the median of the following set of observations? What is the mode?

55, 50, 53, 54, 53, 54, 55, 56, 58, 60, 54, 54, 55, 58, 52, 54, 56, 57, 59

Answer:

Median, 55
Mode, 54

> *When there is an even number of observations, the middle of the list is halfway between two observations. The usual procedure is to call the halfway point the median; for example, for*

1, 5, 6, 9, 11, 12

> *the median is 7.5 (halfway between 6 and 9).*

26 The *mean* is the value obtained by adding all the measurements and dividing by the number of measurements. For example,

1.0

2.0

3.0

6.0

6.0

7.0

9.0
_____

34.0

$34.0 \div 7 = 4.9$

For this series of observations, find the mean, median, and mode.

400, 600, 800, 800, 850, 850, 850, 850, 850, 950, 1000

Answer:

Mean, 800 (= 8800/11)
Median, 850
Mode, 850

27 Let's look at the formula for the mean. It looks a little alarming at first, but since you have already computed a mean you know that the process is simple.

The formula for the mean is

$$\mu = \frac{\Sigma x}{n}$$

In this formula

μ (Greek letter mu) stands for the mean

x stands for each of the individual observations

Σ (summation sign) indicates the operation of summing all the values of x

n stands for the number of observations

We can read the formula as "mu equals the summation of x over n." Copy the formula once to get used to it.

Answer:

$$\mu = \frac{\sum x}{n}$$

28 Now let us apply the formula to some numbers. The following are the durations in minutes of a series of telephone calls. You want to know the mean duration of the calls.

1.0, 5.0, 5.3, 6.0, 6.0, 7.0, 7.7, 8.0, 8.1, 8.9

$$\mu = \frac{\sum x}{n}$$

In this problem x has a series of different values. What are they?

Answer: 1.0, 5.0, 5.3, etc., the duration of each call

29 $\sum x$ is the sum of all the values of x. What is $\sum x$ in this case?

Answer: 63.0 (= 1.0 + 5.0 + 5.3 + . . . , etc.)

30 What is n in this case?

Answer: 10 (the number of observations you have for x)

31 What is μ ?

Answer:

$$\mu = \frac{\sum x}{n} = \frac{63.0}{10} = 6.3$$

CENTRAL TENDENCY WITH SPREADSHEET PROGRAMS

Spreadsheet programs provide functions that will automatically find measures of central tendency. In each case, you must enter a formula in a cell. The formula identifies the range that contains the data and the desired measure of central tendency.

Measure of Central Tendency	Microsoft Excel	Lotus 1-2-3
Mean	=AVERAGE(A1:A20)	@avg(A1..A20)
Median	=MEDIAN(A1:A20)	@median(A1..A20)
Mode	=MODE(A1:A20)	not available

 Answer this question for the spreadsheet program you use.

If your data are in cells D15 through D75, and you want to display the mean in cell E5, what formula will you enter in cell E5?

_____

Answer:

If you use Microsoft Excel, =AVERAGE(D15:D75)
If you use Lotus 1-2-3, @avg(D15..D75)

MEASURES OF VARIABILITY

A measure of variability is a way of indicating how dispersed a set of observations is. The range of a distribution—the difference between the largest observation in the distribution and the smallest—is one crude measure of variability. The other measures of variability you will learn in this book are the standard deviation and the variance.

Even if you plan to use a computer to compute standard deviations and variances, we strongly suggest you use a hand calculator or pencil and paper to complete the computations in this section. Working through the process will help you remember the meanings of the formulas.

33   Standard deviation and variance are measures of _____ .

Answer: Variability

34   The basic formula for standard deviation is

$$\sigma = \sqrt{\frac{\Sigma(x - \mu)^2}{n}}$$

where σ (Greek letter sigma) stands for the standard deviation. The other symbols should be familiar to you. If your math is a little rusty, don't panic! We will walk through the formulas step by step. (If you have already seen a somewhat different formula, be patient; we will get to it in a moment.)

Let us apply the formula step-by-step to the following data:
(Refer to the worksheet on the following page.)

Step 1.   Find the mean, μ (see shaded area 1).

Step 2.   Find $(x - \mu)$ for each of the values of x. The parentheses indicate that this step comes first (see shaded area 2).

Step 3.   Find $(x - \mu)^2$ for each of the values of x and then find the sum of all these values of $(x - \mu)^2$ (see shaded area 3).

Step 4.   Divide $\Sigma(x - \mu)^2$ by the number of values of x (see shaded area 4).

Step 5.   Find the square root of $[\Sigma(x - \mu)^2]/n$. (Use a calculator or look it up in a table, such as Table II in the back of the book; see shaded area 5.)

1. x	2. $(x - \mu)$	3. $(x - \mu)^2$
2	−1	+1
2	−1	+1
2	−1	+1
3	0	0
3	0	0
4	+1	+1
4	+1	+1
4	+1	+1
$\sum x = 24$		$\sum (x - \mu)^2$

$$\mu = \frac{\sum x}{n} = \frac{24}{8} = 3$$

4. $\dfrac{\sum (x - \mu)^2}{n} = \dfrac{6}{8} = 0.75$

5. $\sigma = \sqrt{\dfrac{\sum (x - \mu)^2}{n}} = \sqrt{0.75} = 0.866$

Now apply the same procedure to the following new data by completing the steps below:

1. x

1

2

2

2

3

4

4

4

5

$\sum x =$

$\mu = \dfrac{\sum x}{n} =$

2. $(x - \mu)$

3. $(x - \mu)^2$

$\sum(x - \mu)^2 =$

4. $\dfrac{\sum(x - \mu)^2}{n} =$

5. $\sigma = \sqrt{\dfrac{\sum(x - \mu)^2}{n}} =$

Answer:

1.	x	2.	$(x - \mu)$	3.	$(x - \mu)^2$
	1		−2		+4
	2		−1		+1
	2		−1		+1
	2		−1		+1
	3		0		0
	4		+1		+1
	4		+1		+1
	4		+1		+1
	5		+2		+4
	$\sum x = 27$				$\sum (x - \mu) = 14$

$$\mu = \frac{\sum x}{n} = \frac{27}{9} = 3$$

4. $\dfrac{\sum (x - \mu)^2}{n} = \dfrac{14}{9} = 1.556$

5. $\sigma = \sqrt{\dfrac{\sum (x - \mu)^2}{n}} = \sqrt{1.556} = 1.248$

35 The reasons for using this particular formula as the usual measure of variability depend on mathematical considerations that are beyond the scope of this book. Other formulas are possible, but this one happens to be the most useful. To make sure you understand the formula for standard deviation, let us restate it in other terms. First, the expression $(x - \mu)$ is simply the difference between each observation and the _____ .

Answer: Mean

36 Once the differences from the mean are computed, the next step is to _____ .

Answer: Square them

37 The squared differences from the mean are then added up and divided by the number of observations; that is, they are _____ .

Answer: Averaged

38 The final step is to _____ .

Answer: Take the square root

39 "The standard deviation is the square root of the mean squared deviation from the mean." True or false?

Answer: True

40 Compute the standard deviation of this set of numbers:

1, 1, 3, 5, 5

Answer:

x	$(x - \mu)$	$(x - \mu)^2$
1	−2	+4
1	−2	+4
3	0	0
5	+2	+4
5	+2	+4
$\sum x = 15$		$\sum(x - \mu)^2 = 16$

$$\mu = \frac{\sum x}{n} = \frac{15}{5} = 3$$

$$\frac{\sum(x - \mu)^2}{n} = \frac{16}{5} = 3.2$$

$$\sigma = \sqrt{\frac{\sum(x - \mu)^2}{n}} = \sqrt{3.2} = 1.79$$

41 Variance is simply σ^2. If the standard deviation of a distribution is 2, what is its variance?

Answer: 4

42 If the variance of a distribution is 25.00, what is its standard deviation?

Answer: 5.00

43 You compute variance the same way that you compute the standard deviation, except that you skip the last step—taking the square root. So the formula for σ^2 is _____ .

Answer:

$$\sigma^2 = \frac{\sum (x - \mu)^2}{n}$$

44 There is another formula for σ and σ^2 you will often see. It is mathematically equivalent to the one you have already learned, but it is easier to use for computation because it avoids computing $(x - \mu)$ for each observation.

$$\sigma = \sqrt{\frac{\sum x^2 - (\sum x)^2/n}{n}}, \quad \sigma^2 = \frac{\sum x^2 - (\sum x)^2/n}{n}$$

Let us work through this formula once to give you more practice in interpreting statistical formulas. It is not necessary to memorize the formula. The observations are:

1, 2, 2, 2, 3, 4, 4, 4, 5

Read the following description of the steps and complete the table to find σ^2.

Step 1. Find $\sum x$ by adding all the values of x. Then compute $(\sum x)^2$ by squaring $\sum x$ and then divide by the number of values of x to find $(\sum x)^2/n$. Remember, perform the steps in parentheses first.

Step 2. Find x^2 for each value of x and find the sum of all these values of x^2, $\sum x^2$.

Step 3. Compute $\sum x^2 - (\sum x)^2/n$ and then divide the result by n.

Complete the following computation:

1.	x		2.	x^2
	1			1
	2			4
	2			4
	2			4
	3			9
	4			16
	4			16
	4			16
	5			25

$\sum x =$

$\sum x^2 =$

$\dfrac{(\sum x)^2}{n} =$

$\qquad \sum x^2 =$

3. $\sigma^2 = \dfrac{\sum x^2 - (\sum x)^2/n}{n}$

Answer:

1. x
1
2
2
2
3
4
4
4
5
——
$\sum x = 27$

$$(\sum x)^2 = 729$$

$$\frac{(\sum x)^2}{n} = \frac{729}{9} = 81$$

2. x^2
1
4
4
4
9
16
16
16
25
——
$\sum x^2 = 95$

$$3.\ \sigma^2 = \frac{\sum x^2 - (\sum x)^2/n}{n} = \frac{95 - 81}{9} = \frac{14}{9} = 1.56$$

45   It is important to keep the steps in the right order when using statistical formulas.

$(\sum x)^2$ means "find the sum of the values of x, then square it."

$\sum x^2$ means "find the values of _____ , then find _____ ."

Answer:

x^2

The sum

46 $(\Sigma\, x)^2$ means "first _____ , then _____ ."

Answer:

Find the sum
Find the square

47 $\Sigma\, x^2$ means "first _____ , then _____ ."

Answer:

Find the square
Find the sum

48 What does the symbol μ stand for?

Answer: Mean

49 What does the symbol σ^2 stand for?

Answer: Variance

50 What does the symbol σ stand for?

Answer: Standard deviation

51 For the following data, find the median, μ, σ, and σ^2. Do the computations by hand or by using a calculator.

1.0, 1.0, 2.0, 2.0, 3.0, 5.0, 7.0

Answer: The median is the fourth (middle) observation in order of magnitude, 2.0.

x	$(x - \mu)$	$(x - \mu)^2$
1	−2	+4
1	−2	+4
2	−1	+1
2	−1	+1
3	0	0
5	+2	+4
7	+4	+16
$\sum x = 21$		$\sum(x - \mu)^2 = 30$

$$\mu = \frac{\sum x}{n} = \frac{21}{7} = 3$$

$$\sigma^2 = \frac{\sum(x - \mu)^2}{n} = \frac{30}{7} = 4.29$$

$$\sigma = \sqrt{\frac{\sum(x - \mu)^2}{n}} = \sqrt{4.29} = 2.07$$

VARIABILITY WITH SPREADSHEET PROGRAMS

Now that you have done the work of computing σ and σ^2 by hand, you should be very pleased to know that spreadsheet programs also provide functions that will automatically find measures of variability. As with measures of central tendency, you must enter a formula in a cell. The formula identifies the range that contains the data and the desired measure of variability.

Measure of Variability	Microsoft Excel	Lotus 1-2-3
Standard deviation	=STDEVP(A1:A20)	@std(A1..A20)
Variance	=VARP(A1:A20)	@var(A1..A20)
Maximum	=MAX(A1:A20)	@max(A1..A20)
Minimum	=MIN(A1:A20)	@min(A1..A20)
Range	=(MAX(A1:A20)− MIN(A1:A20))	(@max(A1..A20)− @min(A1..A20))

(Warning, there are also formulas for sample-based estimates of standard deviation and variance. You will learn about these in chapter 3.)

52 Answer this question for the spreadsheet program you use.
If your data are in cells D15 through D75, and you want to display the variance in cell E5 and the standard deviation in cell E6, what formula will you enter in cell E5? _____ What formula will you enter in cell E6? _____

Answer:

If you use Microsoft Excel, E5, =VARP(D15:D75); E6, =STEDVP(D15:D75).
If you use Lotus 1-2-3, E5, @var(D15..D75); E6, @std(D15..D75).

SELF-TEST If you have successfully completed this chapter, you can now summarize data in a number of ways. You can:

• Construct a frequency distribution

• Describe a frequency distribution as approximately normal, skewed, or bimodal

• Compute the mean, median, and mode of your data

• Compute the standard deviation and variance of your data

Since you can perform these computations yourself, you also know what the computer has done when it presents you with a report of these numbers.

Now try these review problems. Table I in the back of the book lists any formulas you may need for reference. Table I is perforated so you can tear it out easily, if you wish.

1. Construct a frequency distribution for the following data:

0.1	2.5	2.6	5.1	5.3	6.7	7.1
7.3	7.5	7.5	8.9	9.9	10.1	11.3
11.7	12.5	12.8	14.1	15.0	17.5	18.9
19.8	21.7	24.4	24.9			

2. Sketch a distribution that is skewed to the right.

3. What is the median of the data in question 1?

4. What is the mean of the data in question 1?

5. What is the standard deviation of the following data?

    3, 8, 8, 8, 9, 9, 9, 18

ANSWERS

To review a problem, study the frames indicated after the answer.

1. The intervals used for the following distribution are from 0.0 to but not including 2.50, from 2.50 to but not including 5.00, and so on. You may have used different intervals, but your graph should be similar to this one. *See frames 1 to 10.*

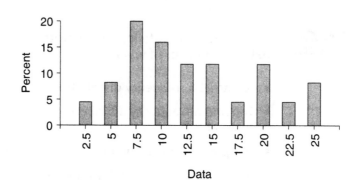

FIGURE 1-23:
Frequency distribution

2.

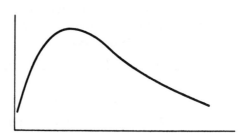

FIGURE 1-24:
Skewed to the right

See frames 12 to 23.

3. Median = 10.1

    See frames 24 to 26.

4. $\mu = \dfrac{\sum x}{n} = \dfrac{285.2}{25} = 11.41$

    See frames 27 to 32.

5. $\sigma = \sqrt{\dfrac{\sum (x-\mu)^2}{n}} = \sqrt{\dfrac{120}{8}} = \sqrt{15} + 3.87$

    See frames 34 to 52.

2 Populations and Samples

The distinction between a *population* and a *sample* is very important in statistics. A *population* includes all possible observations of a particular type. For example, if you measured the height of every tree in the forest, you would be measuring a population. A *sample* includes only some of the observations, but selected in a way that gives every possible observation an equal chance of being chosen. For example, if you randomly chose fifty trees from a forest and measured their heights, you would be measuring a sample from a population.

Most of the techniques you will learn in this book have to do with using information about a sample to draw conclusions about the population, or vice versa. From information about a few randomly selected trees, you can draw some conclusions about the forest. Also, from information about the forest, you can draw some conclusions about a randomly selected group of trees.

The tool you use to go back and forth between samples and population is a mathematical table called a *sampling distribution*. In this chapter, you will get a brief view of how sampling distributions are mathematically derived. You will also learn how to use two of the most useful sampling distribution tables to draw conclusions about a sample on the basis of information about the population.

When you have completed this chapter, you will be able to:

• Distinguish between a population and a sample

• Use a sampling distribution

• Predict a sample proportion, using a binomial probability table

• Predict a sample mean using a normal probability table

POPULATIONS AND SAMPLES

A population is all conceivable observations of a particular type. A sample is a limited number of observations from a population, chosen in a way that allows every possible observation an equal chance of occurring. Using statistics, it is possible to make statements about what a population is probably like on the basis of information from a sample. It is also possible to make statements about what samples will probably be like on the basis of information about the population.

1 In some cases, we use statistical techniques to go from information about a sample to information about a population. In other cases, we do just the opposite and use information about a population to draw conclusions about the probable characteristics of a sample. For example, a researcher is interested in the learning ability of rats that receive a particular diet (she wants to compare the effect of this diet with that of another diet). There is no clear theoretical limit to the number of rats she could raise on the special diet, but for practical reasons she settles for measuring the learning ability of fifty rats. In this example the fifty rats are the _____ .

Answer: Sample

2 All rats who receive the special diet are the _____ .

Answer: Population

3 The researcher will

(a) Use information about the sample to make statements about what the population is probably like

(b) Use information about the population to make statements about what the sample is probably like

Answer: (a)

4 On the basis of certain physical laws, an engineer has determined that one out of every twenty-seven components of a given type will be defective. He wants to determine the probability of finding two or more defective components in a particular batch of five. In this case the population is _____ .

Answer: All components

5 The sample is _____ .

Answer: The batch of five

6 The engineer will

(a) Use information about the sample to make statements about what the population is probably like

(b) Use information about the population to make statements about what the sample is probably like

Answer: (b)

7 Assume that the United States Census Bureau asks every inhabitant of the United States for his age and asks every 100th inhabitant for additional information about his education. The frequency distribution of ages obtained from these observations would be a (sample/population) distribution.

Answer: Population (assuming no one was missed in the census)

8 The frequency distribution of education levels would be a (sample/population) distribution.

Answer: Sample

9 In order to find out which of the four supermarkets in your neighborhood has the best prices, you compile a typical shopping list and price the items on the list at all four stores. To decide whether the figures you obtain are samples or populations, ask yourself, "Are the observations telling the complete story or am I just assuming that other observations will be similar?" Are these shopping list figures populations or samples?

Answer: Samples. The populations are all prices at each store; you are assuming that on other days and with other shopping lists you will obtain similar results.

10 A teacher wants to know the ages of the children in his class. He looks up the age of each child in the school records. The information he obtains is a (sample/population).

Answer: Population

11 A number used to summarize a population distribution is called a *parameter.* A similar number used to describe a sample distribution is called a *statistic.* If you are studying the population of the United States, the mean age of all inhabitants of the United States is a (parameter/statistic).

Answer: Parameter

12 A researcher wants to estimate the average number of ladybugs per acre in Nebraska cornfields. To do so he counts the number of ladybugs in a large number of randomly selected one-acre plots. The mean number of ladybugs per acre in his sample is a (parameter/statistic).

Answer: Statistic

13 The mean number of ladybugs per acre in Nebraska cornfields is a (parameter/ statistic).

Answer: Parameter. It describes the state of affairs in the population of all Nebraska cornfields, although, of course, we have no way of counting every ladybug in every acre.

14 The mean of a sample distribution is a _____ .

Answer: Statistic

15 The mean of a population distribution is a _____ .

Answer: Parameter

16   The standard deviation of a population distribution is a _____ .

Answer: Parameter

17   The standard deviation of a sample distribution is a _____ .

Answer: Statistic

As we noted earlier, a sample should be chosen in such a way that every possible observation has an equal chance of occurring. However, this requirement is often very difficult to meet. Researchers in various fields have developed specialized techniques to assure random selection of samples. You will probably be on safe ground if you learn and use the accepted methods of sampling for your field, but you should be aware that the proper selection of samples is one of the most difficult areas in the application of statistics. As you work through this book, you will find notes pointing out some of these difficulties. For a deeper understanding of these issues, you will need to study experimental design.

SAMPLING DISTRIBUTIONS

The idea of a sampling distribution is the key to our ability to reason back and forth from populations to samples and from samples to populations.

18   Let us develop a simple example of a sampling distribution. Take a coin out. You are interested in the proportion of times that the coin will come up heads when you flip it. We will call the event we are interested in a "success." Flip the coin twice and note how often it comes up heads. Write your results here:

first toss _____

second toss _____

The results of your two tosses are a (population/sample).

Answer: Sample

19 What is the population?

Answer: The result of all possible tosses. Note that in this case the population is unlimited.

20 Look back at your sample. What proportion (p) of the time did the coin come up heads? p = number of successes (heads)/ number of tosses = _____ .

Answer:

If your coin came up heads twice, $p = 1.0$
If your coin came up heads once, $p = 0.5$ (or 1/2)
If your coin came up tails both times, $p = 0.0$

Note that p must always be between 1.0 and 0.0.

21 The value for p you just computed is a (parameter/statistic).

Answer: Statistic. It describes a sample. To distinguish between the parameter and the statistic, we use the capital letter P to identify the proportion of an entire population and the lowercase p to identify the proportion of a sample.

22 Think about your general experience with flipping coins. If you could perform a similar computation for the population, what value for P would you obtain?

$P =$

Answer: $P = 0.5$. Assuming that you are not using trick coins, the coin should come up heads half the time and tails half the time.

23 Take another sample of two tosses and compute p.

$p =$

Answer:

If your coin came up heads twice, $p = 1.0$
If it came up heads once, $p = 0.5$
If it came up tails both times, $p = 0.0$

24 Take three more samples. Remember a sample consists of two tosses. You have now taken five samples of two tosses each. Summarize the results of your sampling below:

Sample p

1

2

3

4

5

Will the statistic p always equal 0.5?

Answer: No. In fact it is unlikely that you found a value of 0.5 for all five samples.

25 If you continue to take a large number of samples of two and you average the values of p that you obtain from each sample, what do you think will be the average value of p?

Answer: The average value of p will be approximately 0.5.

26 Can you construct a frequency distribution for the values of the statistic p that you obtain from a number of samples?

Answer: Yes

27 Draw a frequency distribution graph for the five values of p you have obtained from different samples.

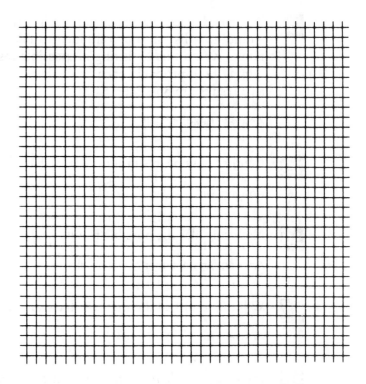

FIGURE 2-1:
Draw a frequency
distribution of *p*

Answer: Your frequency distribution should look something like the following graphs. The mode will probably be 0.5, but not necessarily.

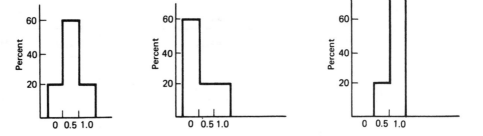

FIGURE 2-2:
Possible frequency
distributions of *p*

28 We can deduce mathematically what the frequency distribution of p would look like if you continued to take samples of two tosses indefinitely. We do this by listing all possible results of two tosses and determining how probable each result is. First consider the first toss. What two results are possible?

Answer: Heads or tails

29 Is either possible result more likely than the other?

Answer: No

30 Now consider the possible results of the second toss. What are they?

Answer: Again, either heads or tails

31 Will the results of the first toss have any influence on the second toss?

Answer: No

32 Is either possible result of the second toss more likely than the other?

Answer: No

33 We can summarize the possible results of the two tosses in the following table:

First Toss	Second Toss
Heads	Heads
	Tails
Tails	Heads
	Tails

As you can see, there are four possible results of the two tosses:

(a) Heads on the first toss and heads on the second toss

(b) _____

(c) _____

(d) _____

Answer:

(b) Heads on the first toss and tails on the second toss

(c) Tails on the first toss and heads on the second toss

(d) Tails on the first toss and tails on the second toss

34 Is any of these results more likely than any of the others?

Answer: No, each of the four results is equally likely.

35 The possible results of the two tosses are

(a) Heads   Heads   _____

(b) Heads   Tails   _____

(c) Tails   Heads   _____

(d) Tails   Tails   _____

Next to each result write the corresponding value of the statistic p.

$$p = \text{number of successes/number of tosses}$$

(That is, number of heads/tosses.)

Answer:

(a) $p = 1.0$

(b) $p = 0.5$

(c) $p = 0.5$

(d) $p = 0.0$

36 Draw a frequency distribution for the statistic p based on this mathematical analysis.

FIGURE 2-3:

Draw a theoretical frequency distribution of p

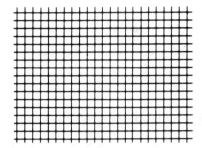

Answer: Your distribution should look like this:

FIGURE 2-4:

Theoretical frequency distribution of p

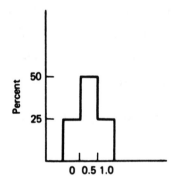

37 If you continue to take samples of two and compute the statistic p, you will find that the frequency distribution of the values of p comes closer and closer to matching the theoretical distribution you have just deduced. If you feel so inclined, take another fifteen samples of two tosses and compute values of p. Compare the actual distribution of the twenty values of p with the theoretical distribution shown by the dark line on the graph below. (If you have a spreadsheet program available, this would be a good opportunity to practice using it to do a frequency distribution. See chapter 1, frame 11.)

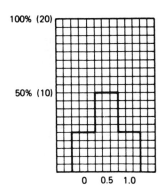

100% (20)

50% (10)

0   0.5   1.0

FIGURE 2-5:

Actual versus theoretical frequency distribution of *p*

38 The distribution of a statistic is called a sampling distribution. The distributions of *p* you have been working with are _____ distributions.

Answer: Sampling

39 Look at the distributions below (a, b, and c).

Which is a sampling distribution? _____

Which is a sample distribution? _____

Which is a population distribution? _____

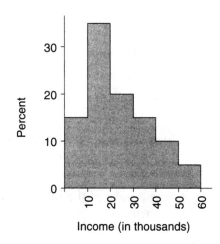

FIGURE 2-6a:

Distribution of incomes in a random sample of 1,000 households

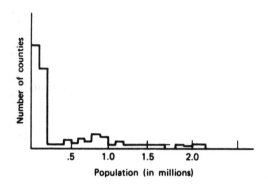

FIGURE 2-6b:
Distribution of county populations in 1990 census data

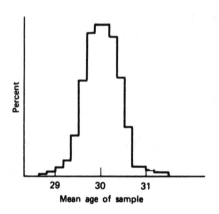

FIGURE 2-6c:
Distribution of mean age in 1,000 samples of 20 individuals each, drawn from a population of mean age 30

Answer:

Sampling distribution	(c) This is the distribution of a statistic (the mean ages of 1,000 samples)
Sample distribution	(a) This is the distribution of observations in a sample (1,000 random households)
Population distribution	(b) This is the distribution of all possible observations in the population (all counties in the United States)

40 The distribution of the individual observations in a sample is called a _____ distribution.

Answer: Sample

41 The distribution of all conceivable observations is called a _____ distribution.

Answer: Population

42 The distribution of a statistic is called a _____ distribution.

Answer: Sampling

43 The median weight of the 20,000 members of a national health club is 175 lb. We would like to know what the probability is that three members, selected at random, all weigh 175 lb or more. To find out, we can use a method similar to the one we used to analyze the coin-tossing problem. First, in the population with which we are dealing (members of the club), what proportion weighs 175 lb or more?

$$P = \_\_\_\_$$

Answer: 0.5. Remember, the median is the point in the distribution at which half the measurements are higher and half lower. So, if half the population weighs 175 lb or more,

$$P = \frac{10,000}{20,000} = 0.5$$

44 We can use this information to construct a theoretical sampling distribution for samples of three drawn from this population. What are the possible outcomes of selecting the first member at random?

Answer:

Equal to or more than 175 lb
Less than 175 lb

45 Is either more likely than the other?

Answer: No

46 If the population is large, will the first choice have any effect on the next choice?

Answer: No

47   Draw a table like the one in frame 33 to show all possible results of selecting three members at random.

First Selection	Second Selection	Third Selection

Answer:

First Selection	Second Selection	Third Selection
175 or over	175 or over	175 or over
		Under 175
	Under 175	175 or over
		Under 175
Under 175	175 or over	175 or over
		Under 175
	Under 175	175 or over
		Under 175

48   Using the table, list all possible results of the three choices and compute the proportion that weighs 175 lb or more for each possible result. That is, count the number of "successes" and divide by the number of trials. (Note: "Over" includes 175 lb and more.)

	Result	*p*
(a)	Over over over	
(b)	Over over under	
(c)	Over under over	
(d)	Over under under	
(e)	Under over over	
(f)	Under over under	
(g)	Under under over	
(h)	Under under under	

Answer:

	Result	*p*
(a)	Over over over	$\frac{3}{3}$ or 1.00
(b)	Over over under	$\frac{2}{3}$ or 0.67
(c)	Over under over	$\frac{2}{3}$ or 0.67
(d)	Over under under	$\frac{1}{3}$ or 0.33
(e)	Under over over	$\frac{2}{3}$ or 0.67
(f)	Under over under	$\frac{1}{3}$ or 0.33
(g)	Under under over	$\frac{1}{3}$ or 0.33
(h)	Under under under	$\frac{0}{3}$ or 0.00

49 Draw a sampling distribution for *p* under these conditions.

Answer:

FIGURE 2-7:

Sampling distribution of *p*

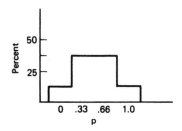

50 We are interested in how often we could expect to draw a sample of three members and find that all three weigh 175 lb or more. To what value of the statistic p would this situation correspond?

Answer: $p = 1.0$

51 Look at your sampling distribution for p. What percent of the time would you expect to draw a sample that gave you a p of 1.0?

Answer: 12.5 percent of the time

THE BINOMIAL PROBABILITY DISTRIBUTION

The binomial probability distribution describes the sampling distribution of the statistic p for many possible values of P and many possible sample sizes.

52 Using the same sort of reasoning you have just applied to the coin tossing and weight problems, it is possible to work out sampling distributions for p not only for various sizes of sample but also for cases in which P is some value other than 0.5; for example, suppose you know that in a large population one out of four is a college graduate. What is P in this case?

Answer: $P = 0.25$ or $1/4$

53

P														
n	x	0.050	0.100	0.200	0.25	0.300	0.400	0.500	0.600	0.700	0.750	0.800	0.900	0.950
2	0	0.902	0.810	0.640	0.563	0.490	0.360	0.250	0.160	0.090	0.063	0.040	0.010	0.002
	1	0.095	0.180	0.320	0.375	0.420	0.480	0.500	0.480	0.420	0.375	0.320	0.180	0.095
	2	0.002	0.010	0.040	0.063	0.090	0.160	0.250	0.360	0.490	0.563	0.640	0.810	0.902
3	0	0.857	0.729	0.512	0.422	0.343	0.216	0.125	0.064	0.027	0.016	0.008	0.001	0.000
	1	0.135	0.243	0.384	0.422	0.441	0.432	0.375	0.288	0.189	0.141	0.096	0.027	0.007
	2	0.007	0.027	0.096	0.141	0.189	0.288	0.375	0.432	0.441	0.422	0.384	0.243	0.135
	3	0.000	0.001	0.008	0.016	0.027	0.064	0.125	0.216	0.343	0.422	0.512	0.729	0.857
4	0	0.815	0.656	0.410	0.316	0.240	0.130	0.062	0.026	0.008	0.004	0.002	0.000	0.000
	1	0.171	0.292	0.410	0.422	0.412	0.346	0.250	0.154	0.076	0.047	0.026	0.004	0.000
	2	0.014	0.049	0.154	0.211	0.265	0.346	0.375	0.346	0.265	0.211	0.154	0.049	0.014
	3	0.000	0.004	0.026	0.047	0.076	0.154	0.250	0.346	0.412	0.422	0.410	0.292	0.171
	4	0.000	0.000	0.002	0.004	0.008	0.026	0.062	0.130	0.240	0.316	0.410	0.656	0.815
5	0	0.774	0.590	0.328	0.237	0.168	0.078	0.031	0.010	0.002	0.001	0.000	0.000	0.000
	1	0.204	0.328	0.410	0.396	0.360	0.259	0.156	0.077	0.028	0.015	0.006	0.000	0.000
	2	0.021	0.073	0.205	0.264	0.309	0.346	0.312	0.230	0.132	0.088	0.051	0.008	0.001
	3	0.001	0.008	0.051	0.088	0.132	0.230	0.312	0.346	0.309	0.274	0.205	0.073	0.021
	4	0.000	0.000	0.006	0.015	0.028	0.077	0.156	0.259	0.360	0.396	0.410	0.328	0.204
	5								0.078		0.237	0.328	0.590	0.774

FIGURE 2-8:

Table of binomial probabilities

These sampling distributions may be found listed in a table called a *binomial probability* table. (As you will see in the next section, a spreadsheet program can also calculate these probabilities. But let us begin by understanding the process with a table.) A small excerpt from such a table is given above. To use the table you must know what P is and what the sample size is. The usual symbol for sample size is n. The sampling distribution is given in terms of the number of "successes" rather than values of the statistic p; for example, in the case of tossing coins, P was 0.5 and the sample size n was 2. The appropriate sampling distribution is circled in the table. The number of "successes"—that is, the number of times the coin comes up heads—is indicated by x. According to the circled portion of the table, how frequently will there be two successes in a sample of two?

Answer: 0.25; that is, 25 percent of the time

54 For the weight problem, what was *P*? Look back at frame 43 if you do not recall the problem.

Answer: $P = 0.5$

55 What was *n*?

Answer: $n = 3$ (*n* is the sample size)

56 Circle the appropriate sampling distribution in the table in frame 53.

Answer:

n	x	0.050	0.100	0.200	0.250	0.300	0.400	0.500	0.600	0.700	0.750	0.800	0.900	0.950
								P						
2	0	0.902	0.810	0.640	0.563	0.490	0.360	0.250	0.160	0.090	0.063	0.040	0.010	0.002
	1	0.095	0.180	0.320	0.375	0.420	0.480	0.500	0.480	0.420	0.375	0.320	0.180	0.095
	2	0.002	0.010	0.040	0.063	0.090	0.160	0.250	0.360	0.490	0.563	0.640	0.810	0.902
3	0	0.857	0.729	0.512	0.422	0.343	0.216	0.125	0.064	0.027	0.016	0.008	0.001	0.000
	1	0.135	0.243	0.384	0.422	0.441	0.432	0.375	0.288	0.189	0.141	0.096	0.027	0.007
	2	0.007	0.027	0.096	0.141	0.189	0.288	0.375	0.432	0.441	0.422	0.384	0.243	0.135
	3	0.000	0.001	0.008	0.016	0.027	0.064	0.125	0.216	0.343	0.422	0.512	0.729	0.857
4	0	0.815	0.656	0.410	0.316	0.240	0.130	0.062	0.026	0.008	0.004	0.002	0.000	0.000
	1	0.171	0.292	0.410	0.422	0.412	0.346	0.250	0.154	0.076	0.047	0.026	0.004	0.000
	2	0.014	0.049	0.154	0.211	0.265	0.346	0.375	0.346	0.265	0.211	0.154	0.049	0.014
	3	0.000	0.004	0.026	0.047	0.076	0.154	0.250	0.346	0.412	0.422	0.410	0.292	0.171
	4	0.000	0.000	0.002	0.004	0.008	0.026	0.062	0.130	0.240	0.316	0.410	0.656	0.815
5	0	0.000	0.390	0.390	0.237	0.168	0.078	0.031	0.010	0.007	0.001			
			0.008	0.000	0.000									

FIGURE 2-9:
Table of binomial probabilities

57 You are told that in the population of all doctors nine out of ten recommend Potter's Pills. Assuming this is true, you want to know the probability of choosing two doctors at random and finding that neither of them recommends Potter's Pills. What are *P* and *n* for this case?

Answer:

$P = 0.9$ (9/10)
$n = 2.0$

58 What value of x corresponds to the case in which neither of the two doctors recommends Potter's Pills? (A doctor who recommended Potter's Pills would constitute a "success.")

Answer: $x = 0$

59 How likely is it that neither doctor will recommend Potter's Pills? Use the table in frame 53 to obtain your answer.

Answer: The probability of this event is 0.010; that is, it should happen 1 percent of the time.

60 Let us make the problem more complicated. Suppose you plan to choose four doctors at random and you want to know the probability that no more than two will recommend Potter's Pills. To come up with an answer you must take into account the case in which none of them recommends Potter's Pills (zero "successes"), the case in which only one recommends Potter's Pills and three do not (one "success"), and the case in which two recommend Potter's Pills and two do not (two "successes"). You will find the probability of each of these cases in the sampling distribution; then you will add the probabilities for each of these cases. What are P and n for this problem? What is the probability that one of these three situations will exist?

Answer: $P = 0.9$, $n = 4$. The probability that x is less than or equal to 2 is 0.053. To find this answer you had to add the probabilities of the three situations of interest. The probability that $x = 0$ is too small to enter in the table. The probability that $x = 1$ is 0.004, and the probability that $x = 2$ is 0.049.

61 The total of all the probabilities in a sampling distribution is always 1.0. If the probability of no, one, or two recommendations for Potter's Pills is 0.053, what is the probability of three or four recommendations?

Answer: 0.947 (= 1.000 − 0.053)

62 Five percent of the population of the United States possesses a particular genetic trait. You want to know the chances of finding at least one person who has this trait in a random sample of fifteen. Use table III in the back of the book.

Answer: $p = 0.05$, $n = 15$. The probability is 0.537, or about 54 percent. The easiest way to find the answer is to find the probability of $x = 0$, that is, the probability that no one in the sample will have the trait. The probability that no one will have the trait is 0.463; therefore the probability that at least one person will have the trait is $1.000 - 0.463 = 0.537$.

You could also have added up the probabilities for $x = 1$, $x = 2$, $x = 3$, and so on. The sum of these probabilities, using the table of binomial probabilities (Figure 2-8), is 0.538. Because the table entries are rounded off, you will find occasional minor discrepancies like this.

63 When you use the binomial probability table, you are making an important assumption. You are assuming that the observations in your sample are random and independent; that is, you are assuming that every observation in the population has an equal chance of being chosen at any step in the sampling process. Suppose you are going to choose three people out of a population of six that consists of three men and three women. You choose to select them one at a time. On the first draw you choose a man. Does this affect the chances of your choosing a man on the second draw?

Answer: Yes, because there are now substantially fewer men in the remaining population. For the first draw, $P = 3/6 = 0.5$; for the second draw, $P = 2/5 = 0.4$.

64 You have a jar containing six marbles, three red and three blue. You draw one marble at random and find that it is blue. You record this observation and replace the marble in the jar. Has your first observation affected the chances of choosing a blue marble on the second draw?

Answer: No, because you replaced the blue marble so the population remains the same.

65 Can you use the binomial probability table to analyze the marble problem?

Answer: Yes. The observations are independent. One observation does not affect the probability of the others.

66 Can you use the binomial probability table to analyze the problem in frame 63?

Answer: No. The observations are not independent.

> *Procedures for dealing with this problem exist, but they are beyond the scope of this book.*

67 As a rule of thumb, if the population is at least twenty times as large as the sample, you can disregard this effect of not being able to include one individual in your sample twice. If you are choosing a sample of six out of a population of 500, can you use the binomial probability table?

Answer: Yes

68 If you are choosing a sample of six out of a population of seventy-five, can you use the binomial probability table?

Answer: No

69 You recall that if nine out of ten doctors recommend Potter's Pills the probability of selecting two doctors at random who do not recommend them is 0.010. You go to a nearby medical school and ask the first two doctors you meet whether they recommend Potter's Pills. Have all members of the population an equal chance of being chosen?

Answer: No. The chance of choosing doctors connected with that medical school is much greater than the chance of choosing other doctors.

70 Can you say that the probability of two "no" answers is 0.010?

Answer: No

71 What assumption do you make when you use the binomial probability table?

Answer: All members of the population have an equal chance of being selected at any point in the sampling procedure.

BINOMIAL PROBABILITIES WITH SPREADSHEET PROGRAMS

If you are using a spreadsheet program, you do not need to refer to a table to look up binomial probabilities. Spreadsheet programs provide you with functions that calculate binomial probabilities.

Microsoft Excel	Lotus 1-2-3
=BINOMDIST(x, n, P, type)	@binomial(n, x, P, type)
type FALSE: exactly x successes	*type 0: exactly x successes*
type TRUE: at most x successes	*type 1: at most x successes*
	type 2: at least x successes

Notice that these formulas give you a choice of type. You may ask the spreadsheet to tell you the probability of exactly x successes, or you may ask the spreadsheet to add up probabilities and tell you the probablility of at most x successes. (Or, in the case of Lotus 1-2-3, at least x successes.)

72 For example, if nine out of ten doctors recommend Potter's Pills, and you want to know the probability of finding *exactly two* who recommend Potter's in a sample of four tries, $x = 2$, $n = 4$, $P = 0.9$.

Write the formula you would enter into a cell of your spreadsheet to have your program calculate the appropriate probability value. (Answer the question for the spreadsheet program you use.) _____

Answer:

If you use Microsoft Excel, =BINOMDIST(2, 4, 0.9, FALSE)
If you use Lotus 1-2-3, @binomial(4, 2, 0.9, 0)

73 If you want to find the probability of at most two recommendations in four tries (the probability of 0, 1, or 2 successes), you can ask the spreadsheet to do the work of adding up the cumulative probabilities by asking for a cumulative calculation.

Write the formula you would enter into a cell of your spreadsheet to have your program calculate the probability of at most 2 successes. (Answer the question for the spreadsheet program you use.) _____

Answer:

If you use Microsoft Excel, =BINOMDIST(2, 4, 0.9, TRUE)
If you use Lotus 1-2-3, @binomial(4, 2, 0.9, 1)

74 You have a large jar of mixed red and blue marbles, with an equal number of red and blue.

As a statistical experiment, you begin drawing marbles from the jar. Each time you draw a marble from the jar, you record the result in a spreadsheet by entering a "1" if the marble is red and a "0" if the marble is blue; then you return the marble to the jar and draw again.

Cell B75 contains a count of the number of times you have drawn a marble from the jar. Cell B76 contains a count of the number of red marbles you have drawn. You want to determine the probability of this exact result. Write a formula you can enter in your spreadsheet to calculate this probability. (Answer the question for the spreadsheet program you use.) _____

Answer:

If you use Microsoft Excel, =BINOMDIST(B76, B75, 0.5, FALSE)
If you use Lotus 1-2-3, @binomial(B75, B76, 0.5, 0)

THE NORMAL DISTRIBUTION

It is possible to deduce the sampling distributions of a number of other statistics mathematically. This book will not attempt to explain the mathematical reasoning involved in deducing these sampling distributions. Instead, our focus will be on how and why sampling distributions are useful in drawing conclusions about populations and samples.

75 One of the most useful sampling distributions is the one for the mean of a large sample of measurements. This distribution is called the normal distribution. Before explaining it, let us review the symbols we will be using.

μ is the symbol for the _____ of the (sample/population).

σ is the symbol for the _____ of the (sample/population).

Answer:

Mean of the population
Standard deviation of the population

76 What does *n* represent?

Answer: Size of the sample

77 μ is a (parameter/statistic)?

Answer: Parameter

78 The sample mean is a (parameter/statistic)?

Answer: Statistic

79 We represent the sample mean by the symbol $\bar{x}$; for example, if the mean height of all women in the United States is 5 ft. 5 in., and a sample of ten women has a mean height of 5 ft. 9 in.,

$$\mu = \underline{\hspace{1cm}}$$
$$\bar{x} = \underline{\hspace{1cm}}$$

Answer:

μ = 5 ft. 5 in.
$\bar{x}$ = 5 ft. 9 in.

80 If we take a large number of samples and compute an $\bar{x}$ for each sample, we can also compute a mean as well as a standard deviation for the values of $\bar{x}$. We denote these values $\mu_{\bar{x}}$ and $\sigma_{\bar{x}}$, respectively. Complete the following table:

	Sample	Sampling Distribution of Mean	Population
Mean			μ
Standard deviation*	s		

*You will learn more about s in chapter 3.

Answer:

	Sample	Sampling Distribution of Mean	Population
Mean	$\bar{x}$	$\mu_{\bar{x}}$	μ
Standard deviation	s	$\sigma_{\bar{x}}$	σ

81

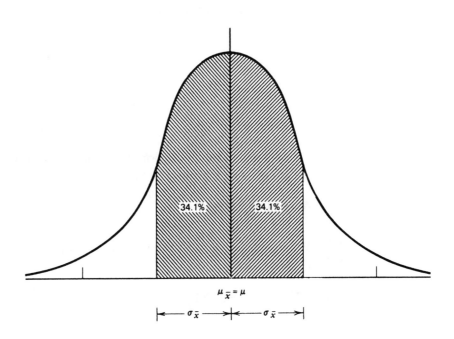

FIGURE 2-10:

The normal distribution

According to what is called the central limit theorem, the sampling distribution of $\bar{x}$ tends to have the shape of the normal distribution curve shown above, with

$$\mu_{\bar{x}} = \mu \text{ and } \sigma_{\bar{x}} = \frac{\sigma}{\sqrt{n}}$$

(If you would like to know the mathematical formula that creates the normal distribution curve, or to see the proof that it represents the sampling distribution of $\bar{x}$, consult an advanced statistics text.)

When the sample size n is thirty or more, the sampling distribution becomes almost exactly like the illustration above. Suppose we took a great number of samples of 36 individuals from a population with a mean age of 40 and a standard deviation of 6 years. What would the mean of the sampling distribution be?

$$\mu_{\bar{x}} = \text{\_\_\_\_\_} = \text{\_\_\_\_\_}$$

Answer:

$$\mu_{\bar{x}} = \mu = 40$$

82 What would be the standard deviation of the sampling distribution?

$$\sigma_{\bar{x}} = \text{\_\_\_\_\_}$$

Answer:

$$\sigma_{\bar{x}} = \frac{\sigma}{\sqrt{n}} = \frac{6}{\sqrt{36}} = 1$$

83 According to the illustration, what percent of the time would you expect the mean of one sample to be between 40 and 41?

Answer: 34.1 percent

84 What percent of the time would you expect the mean of one sample to be between 39 and 41?

Answer: 68.2 percent

85

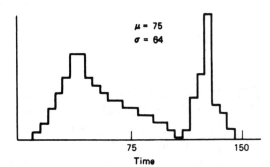

FIGURE 2-11:

Solution times for a population of 6,000

According to the central limit theorem, the shape of the sampling distribution of $\bar{x}$ will be a normal curve, no matter what the shape of the population distribution. The illustration above shows the population distribution for time to solve a particular puzzle for 6,000 individuals. A large number of samples of 49 individuals each are selected, and $\bar{x}$ is computed for each sample. Which of the illustrations below, (a), (b), or (c), will most resemble the sampling distribution of $\bar{x}$?

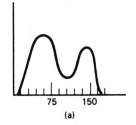

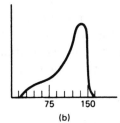

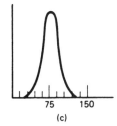

FIGURE 2-12:

Sampling distribution

Answer: (c). No matter what the population distribution or the sample distributions look like, the sampling distribution of the statistic $\bar{x}$ will always have this bell shape.

86 Remember that

$$\sigma_{\bar{x}} = \frac{\sigma}{\sqrt{n}}.$$

What will the mean of the sampling distribution be? What will its standard deviation be?

Answer:

$$\mu_{\bar{x}} = \mu = 75$$

$$\sigma_{\bar{x}} = \frac{\sigma}{\sqrt{n}} = \frac{64}{7} = 9.14$$

87 The shape of the sampling distribution of $\bar{x}$ is called a normal distribution. In this normal distribution 34.1 percent of the observations always lie between $\mu_{\bar{x}}$ and $\mu_{\bar{x}} + \sigma_{\bar{x}}$, 34.1 percent of the observations lie between $\mu_{\bar{x}} - \sigma_{\bar{x}}$ and $\mu_{\bar{x}}$, 47.7 percent of the observations lie between $\mu_{\bar{x}}$ and $\mu_{\bar{x}} + 2\sigma_{\bar{x}}$, and so on. The exact shape of the distribution is completely defined by the two parameters _____ and _____ .

Answer: $\mu_{\bar{x}}$ and $\sigma_{\bar{x}}$

88 If we know how many standard deviations away from the mean an observation is, we can tell how likely it is to occur in random sampling. For this reason it is convenient to convert our measurements into z scores. A z score is simply the number of standard deviations away from the mean a particular measurement is located; for example, a score one standard deviation above the mean has a z score of +1; a score one and one-half standard deviations below the mean has a z score of –1.5.

If a distribution has a mean of 15 and a standard deviation of 2, what z score corresponds to a raw score of 19? What z score corresponds to a raw score of 14?

Answer:

z = +2; that is, two standard deviations above the mean
z = –0.5; that is, one-half standard deviation below the mean

89 The formula for converting any measurement to a z score is

$$z = \frac{\bar{x} - \mu}{\sigma_{\bar{x}}}$$

If $\mu = 15$ and $\sigma_{\bar{x}} = 2$, what z score corresponds to a raw score of 20?

Answer: $z = \frac{20 - 15}{2} = 2.5$

90 The table that gives the probabilities for a normal frequency distribution is usually titled "areas under the normal curve" and is organized in terms of z scores. If you want to know the probability of a particular $\bar{x}$ from a known population, you must first compute a _____ .

Answer: z score

91

An entry in the table is the proportion under the entire curve that is between $z = 0$ and a positive value of z. Areas for negative values of z are obtained by symmetry.

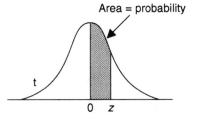

Area = probability

			Second decimal place of z							
z	.00	.01	.02	.03	.04	.05	.06	.07	.08	.09
0.0	.0000	.0040	.0080	.0120	.0160	.0199	.0239	.0279	.0319	.0359
0.1	.0398	.0438	.0478	.0517	.0557	.0596	.0636	.0675	.0714	.0753
0.2	.0793	.0832	.0871	.1910	.0948	.0987	.1026	.1064	.1103	.1141
0.3	.1179	.1217	.1255	.1293	.1331	.1368	.1406	.1443	.1480	.1517
0.4	.1554	.1591	.1628	.1664	.1700	.1736	.1772	.1808	.1844	.1879
0.5	.1915	.1950	.1985	.2019	.2054	.2088	.2123	.2157	.2190	.2224
0.6	.2257	.2291	.2324	.2357	.2389	.2422	.2454	.2486	.2517	.2549
0.7	.2580	.2611	.2642	.2673	.2703	.2734	.2764	.2794	.2823	.2852
0.8	.2881	.2910	.2939	.2967	.2995	.3023	.3051	.3078	.3106	.3133
0.9	.3159	.3186	.3212	.3238	.3264	.3289	.3315	.3340	.3365	.3389
1.0	.3413	.3438	.3461	.3485	.3508	.3531	.3554	.3577	.3599	.3621
1.1	.3643	.3665	.3686	.3708	.3729	.3749	.3770	.3790	.3810	.3830
1.2	.3849	.3869	.3888	.3907	.3925	.3944	.3962	.3980	.3997	.4015
1.3	.4032	.4049	.4066	.4082	.4099	.4115	.4131	.4147	.4162	.4177
1.4	.4192	.4207	.4222	.4236	.4251	.4265	.4279	.4292	.4306	.4319
1.5	.4332	.4245	.4357	.4370	.2222	.4394	.4406	.2222	.4429	.4441

FIGURE 2-13: Areas under the normal curve

The table of areas under the normal curve gives you the probability of a sample mean between μ and $\mu + z$; for example, if you want to know the probability of a sample mean between μ and $\mu + 1.0\sigma$, you will look in the table under $z = 1.00$. The probability in this case is 0.3413. What is the probability of a sample mean between μ and $\mu + 1.5\sigma$?

Answer: 0.4332

92 A population has a mean of 20 and a standard deviation of 4. You plan to choose a sample of 64 at random. What is the probability of a sample mean between 20 and 21? To answer this question, you must find the mean and standard deviation of the sampling distribution and use them to convert 21 to a z score.

What is the mean of the sampling distribution? What is the standard deviation of the sampling distribution? What z score corresponds to 21?

Answer:

$$\mu_{\bar{x}} = 20$$

$$\sigma_{\bar{x}} = \frac{\sigma}{\sqrt{n}} = \frac{4}{\sqrt{64}} = 0.5$$

$$z = \frac{21 - \mu}{0.5} = \frac{1}{0.5} = 2.0$$

93 Use table IV at the back of the book to find the probability of a sample mean between 20 and 21 for the population above.

Answer: 0.4772

94 Often you will have to add or subtract values you find in the table to answer a question; for example, to find the probability of a sample mean between 19 and 21, you must break this down into the probability of an $\bar{x}$ between 20 and 21 and the probability of an $\bar{x}$ between 20 and 19. The answer is _____ .

Answer: 0.9544 (= 0.4772 + 0.4772)

95 It helps to remember that the total of all probabilities in a sampling distribution is always 1.0000. If the probability of a sample mean between 19 and 21 is 0.9544 (i.e., 95 percent of the time), what is the probability of a sample mean that is not between 19 and 21 (either less than 19 or more than 21)?

Answer: 0.0456 (= 1.000 – 0.9544), that is, 5 percent

96 The total of all probabilities on one side of the mean is 0.5000. Therefore, a sample mean will be above the mean of the sampling distribution half the time. What is the probability of a sample mean greater than 21 for the above population?

Answer: 0.0228 (= 0.5000 – 0.4772)

97 Reference formulas

$$\mu_{\bar{x}} = \mu$$

$$\sigma_{\bar{x}} = \frac{\sigma}{\sqrt{n}}$$

$$z = \frac{\bar{x} - \mu}{\sigma_{\bar{x}}}$$

An international airline has determined that the mean weight of commercial express shipments is 28.5 kg, with a standard deviation of 5 kg. If we take random samples of 100 shipments, how often can we expect to find a sample mean over 30 kg?

Answer:

Population distribution:

$$\mu = 28.5$$

$$\sigma = 5.0$$

Sampling distribution:

$$\mu_{\bar{x}} = 28.5$$

$$\sigma_{\bar{x}} = \frac{5}{\sqrt{100}} = 0.5$$

z score corresponding to 30 kg:

$$z = \frac{30 - 28.5}{0.5} = 3.0$$

Probability of $\bar{x}$ between 28.5 and 30:
(from table IV at the back of the book) 0.4987

Probability of $\bar{x}$ over 30:
0.0013 (= 0.5000 – 0.4987), that is, about once in a thousand times

98 The airline's planes hold 300 passengers and, to the delight of the stock-holders, are always full. Can we consider a planeload of passengers as a random sample of 300 airline passengers? Has every member of the population an equal chance of being selected at each stage of the sampling process?

Answer: No, a planeload is not a random sample. For example, the presence of one member of a group tour or a family on a plane greatly increases the probability of other members of the same group being on the plane.

99 According to the central limit theorem, the shape of the sampling distribution of $\bar{x}$ for large samples tends to be normal. Therefore, $\mu_{\bar{x}}$ = _____ and $\sigma_{\bar{x}}$ = _____ .

Answer:

$$\mu_{\bar{x}} = \mu$$

$$\sigma_{\bar{x}} = \frac{\sigma}{\sqrt{n}}$$

100 The shape of the sampling distribution of $\bar{x}$ for large samples (n greater than 30) is called _____ .

Answer: Normal

101 For large samples (n greater than 30), the sampling distribution of $\bar{x}$ tends to be normal, with mean and standard deviation related to the population parameters as follows:

$$\mu_{\bar{x}} = \mu$$

$$\sigma_{\bar{x}} = \frac{\sigma}{\sqrt{n}}$$

This is a statement of the _____ theorem.

Answer: Central limit

On a separate sheet of paper, state the central limit theorem in your own words.

Answer: Your answer should have included these points:

(a) The theorem has to do with the sampling distribution of the means of large samples (greater than 30).

(b) The shape of this sampling distribution tends to be normal.

(c) The mean of this distribution is equal to the population mean.

(d) The standard deviation of this distribution equals the population standard deviation divided by the square root of the sample size.

NORMAL PROBABILITIES WITH SPREADSHEET PROGRAMS

As you probably expect, spreadsheet programs provide you with functions that calculate normal distribution probabilities. The most commonly used functions give a cumulative probability. That is, they report the probability of a z score less than or equal to the score you use for the test. See figure 2-14.

The spreadsheet result is the proportion under the entire curve that is less than the given value of z. Areas between $z = 0$ and the given value are obtained by subtracting 0.5.

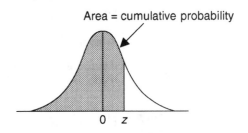

FIGURE 2-14:
Cumulative areas under the normal curve

	Microsoft Excel	Lotus 1-2-3
Cumulative probability	=NORMDIST(x, $\mu_{\bar{x}}$, $\sigma_{\bar{x}}$, TRUE)	@normal(x, $\mu_{\bar{x}}$, $\sigma_{\bar{x}}$, 0)
Cumulative probability (using z score)	=NORMSDIST(z)	@normal(z, 1, 0, 0)
z score	=STANDARDIZE(x, $\mu_{\bar{x}}$, $\sigma_{\bar{x}}$)	$((x - \mu_{\bar{x}})/\sigma_{\bar{x}})$

Notice that the spreadsheet formulas do not require you to compute a z score. Instead, you can enter $\mu_{\bar{x}}$ and $\sigma_{\bar{x}}$ and the spreadsheet program will take care of computing the z score. Also notice that the spreadsheet formulas give a cumulative probability. The spreadsheet tells you the probability of a mean less than or equal to the score you input. Since the total of all probabilities is 1.0 and the probability of a sample mean either above or below the sampling distribution mean is always 0.5, you can calculate other results by addition or subtraction.

103 Remember this problem? An international airline has determined that the mean weight of commercial express shipments is 28.5 kg, with a standard deviation of 5 kg. If we take random samples of 100 shipments, how often can we expect to find a sample mean over 30 kg (cell A1)? You have calculated the necessary inputs in your spreadsheet:

$$\mu_{\bar{x}} = \mu = 28.5 \text{ in cell A2}$$

$$\sigma_{\bar{x}} = \frac{\sigma}{\sqrt{n}} = 0.5 \text{ in cell A3}$$

Write a formula you can enter in your spreadsheet that will return the desired probability. (Answer the question for the spreadsheet program you use.) _____

Answer:

If you use Microsoft Excel, =NORMDIST(A1, A2, A3, TRUE)
If you use Lotus 1-2-3, @normal(A1, A2, A3, 0)

104 Here is another familiar problem. A population has a mean of 20 (cell B2) and a standard deviation of 4 (cell B3). You plan to choose a sample of 64 (cell D4) at random. What is the probability of a sample mean between 20 and 21?

Set up a spreadsheet. (Answer the question for the spreadsheet program you use.)

Answer: To solve this problem, you must subtract the probability of a result less than 20 from the cumulative probability returned by the spreadsheet function. Since 20 is the mean of the sampling distribution, the probability of a sample mean less than 20 is 0.5. See cell D6 of the sample answers.

If you use Microsoft Excel, your spreadsheet will be something like this:

	A	**B**	**C**	**D**
1	*population*		*sample*	
2	mean	20	mean	=B2
3	std	4	std	=B3/SQRT(D4)
4			size	64
5			prob x < 21	=NORMDIST(B2, B3, D4, TRUE)
6			prob 20<x<21	=(D5 – 0.5)

If you use Lotus 1-2-3, your spreadsheet will be something like this:

	A	**B**	**C**	**D**
1	*population*		*sample*	
2	mean	20	mean	=B2
3	std	4	std	=B3/@sqrt(D4)
4			size	64
5			prob x < 21	@normal(B1, B2, D4, 0)
6			prob 20<x<21	D5 – 0.5

SELF-TEST   If you have successfully completed this chapter, you can now use information about the population to make predictions about samples. You can:

- Tell the difference between a population and a sample
- Use a binomial probability table to predict sample proportions, given the population proportion
- Use a normal distribution table to predict sample means, given the mean and standard deviation of the population

Now try these review problems. Table I at the back of the book lists any formulas you may need for reference.

1. Write the term that corresponds to each of the following descriptions:

    (a) A group of observations chosen at random to represent a larger group of possible observations

    (b) All possible observations of a given type

    (c) A number that summarizes a population of observations

    (d) A number that summarizes a sample of observations

2. You have two complete decks of cards with no jokers. You shuffle the cards thoroughly and draw five cards. What is the probability that you will draw four or more clubs?

3. You take only the face cards and aces from a deck of cards and discard the others. Then you shuffle the cards thoroughly and draw two cards. Can you use the binomial probability table to calculate the probability of drawing two clubs? Why or why not?

4. The mean diameter of marbles manufactured at a particular toy factory is 0.850 cm with a standard deviation of 0.010 cm. What is the probability of selecting a random sample of 100 marbles that has a mean diameter greater than 0.851 cm?

ANSWERS

To review a problem, study the frames indicated after the answer.

1. (a) Sample

    (b) Population

    (c) Parameter

    (d) Statistic

    See frames 1 to 17.

2. The population is 104 and the sample size 5, so you may use the binomial probability table (table III in the back of the book), even though the cards are not replaced.

    $P = 0.25$ (there are equal numbers of the four suits).

    $n = 5$ (you will draw five cards).

    $x = 4, 5$ (you are interested in four or more successes).

    From the binomial probability table you find these values: 0.015, 0.001. The probability of drawing four or more clubs is therefore 0.016; it would happen about 1.6 percent of the time. *See frames 52 to 71.*

3. No. Your deck (the population) consists of only 16 cards. Since this is less than 20 times the sample size, you cannot use the binomial probability table unless you replace the card after each individual draw. *See frames 63 to 68.*

4.

$$\mu_{\bar{x}} = 0.85$$

$$\sigma = 0.01$$

$$\sigma_{\bar{x}} = \frac{\sigma}{\sqrt{n}} = \frac{0.01}{10} = 0.001$$

$$z = \frac{\bar{x} - \mu_{\bar{x}}}{\sigma_{\bar{x}}} = \frac{0.851 - 0.850}{0.001} = \frac{0.001}{0.001} = 1.0$$

From the normal probability table (table IV), the probability of $\bar{x}$ between 0.850 and 0.851 is 0.341. The probability of $\bar{x}$ greater than 0.851 is $0.500 - 0.341 = 0.159$, or 15.9 percent. *See frames 75 to 102.*

3 Estimating

Often we need to estimate the characteristics of a population on the basis of information about a sample. For example, you may want to estimate the mean height of all the trees in the forest without having to measure every tree. Common sense suggests that you can get a pretty good idea of the height of all trees in the forest by measuring a random sample of trees. In fact, by using sampling-distribution tables, you can translate the description "pretty good" into a mathematical statement of your level of confidence. This chapter will teach you how to estimate population parameters on the basis of data from a sample and how to establish mathematically defined confidence intervals for these estimates.

When you have completed this chapter, you will be able to:

- Estimate μ, P, and σ

- Establish confidence intervals for P

- Establish confidence intervals for μ

ESTIMATION

Often we use a sample as a basis for estimating the value of population parameters. Even though we know that the sample statistic is not identical to the population parameter, the statistic is our best estimate of the parameter.

1 If the mean weight of 30 cartons of "large" eggs is 30 oz, what would be your best estimate of the mean weight of all cartons of "large" eggs?

Answer: 30 oz

2 In this case you used the statistic _____ as an estimate of the parameter _____ .

Answer:

$\bar{x}$

μ

3 A marketing research organization reports that 18 percent of the television audience was watching *The Return of the Creature from the Black Lagoon* on a given evening. Do you think this proportion is a parameter or a statistic?

Answer: A statistic. This report is based on a sample of television watchers.

4 In this case *p*, a (parameter/statistic), is being used as an estimate of *P*, a (parameter/statistic).

Answer:

Statistic
Parameter

5 When we wish to estimate the standard deviation of a population on the basis of a sample distribution, we encounter a problem. The number that you obtain by applying the formula for σ to the sample data is not a good estimate of the population standard deviation. Instead, a slightly different formula must be used.

> *An explanation would be too long to present here, but with small samples the formula for σ tends to underestimate the standard deviation of the population. That is, on the average, the standard deviation of a sample tends to be smaller than the standard deviation of the population.*

The two formulas are:

Standard deviation of population

$$\sigma = \sqrt{\frac{\sum (x - \mu)^2}{n}}$$

Estimate of population standard deviation based on sample data

$$s = \sqrt{\frac{\sum (x - \bar{x})^2}{n - 1}}$$

The symbol for population standard deviation is _____ .

Answer: σ

6 The symbol for the estimate of population standard deviation is _____ .

Answer: s

7 In these formulas, σ is a (parameter/statistic); s is a (parameter/statistic). The formula for σ contains the expression n where the formula for s contains the expression _____ and contains μ where the formula for s contains _____ .

Answer:

σ is a parameter; s is a statistic

$n - 1$

$\bar{x}$

8 Which formula contains the expression $n - 1$?

Answer:

$$s = \sqrt{\frac{\sum (x - \bar{x})^2}{n - 1}}$$

9 What is the formula for σ?

Answer:

$$\sigma = \sqrt{\frac{\sum (x - \mu)^2}{n}}$$

10 What is the formula for s?

Answer:

$$s = \sqrt{\frac{\sum(x - \bar{x})^2}{n - 1}}$$

11 If you wish to estimate the standard deviation of the weight of all cartons of "large" eggs on the basis of a sample of 30 cartons, which formula would you use (write out the whole formula)?

Answer:

$$s = \sqrt{\frac{\sum(x - \bar{x})^2}{n - 1}}$$

12 If 20 cartons of eggs constitute a population and you wish to compute the standard deviation of this population, which formula would you use (write out the whole formula)?

Answer:

$$\sigma = \sqrt{\frac{\sum(x - \mu)^2}{n}}$$

13 The following numbers represent the number of words in the speaking vocabulary of five children. Describe this group in terms of mean and standard deviation.

100, 100, 300, 400, 600.

Answer:

$$\mu = 300$$

$$\sigma = \sqrt{\frac{180000}{5}} = 189.7$$

Since you are not trying to estimate the parameters of some larger population, you should have used the formula for σ.

14 Five children of a given age in an appropriately selected sample have the following number of words in their speaking vocabularies. What is the best estimate you can make of the mean and standard deviation for all children of that age.

100, 100, 300, 400, 600.

Answer:

$$\bar{x} = 300$$

$$s = \sqrt{\frac{180000}{4}} = 212.1$$

Since this is an estimation problem, you should have used the formula for *s*. (As you will see, with such a small sample these estimates are not very reliable; however, they are the best estimates you can make.)

15 If you compute both *s* and σ for the same data, which will be larger?

Answer: *s*, because the divisor, the number on the bottom of the fraction, is smaller.

16 With a sample size of 1,000, would you expect the difference between σ and *s* to be substantial?

Answer: No. The difference between a given number divided by 1,000 and the same number divided by 999 is not very great.

17 With a sample size of 4, would you expect the difference between σ and *s* to be substantial?

Answer: Yes. There is a substantial difference between *x*/3 and *x*/4.

18 What statistic is used as an estimate of *P*?

Answer: p

19 What statistic is used as an estimate of σ?

Answer: s

20   What statistic is used as an estimate of μ?

Answer: $\bar{x}$

SPREADSHEET FORMULAS FOR ESTIMATES OF VARIABILITY

Just as there are spreadsheet functions that compute σ and σ^2 for you, there are also functions that compute s and s^2.

Estimate of Sample Variability	Microsoft Excel	Lotus 1-2-3
Standard deviation s	=STDEV(A1:A20)	@stds(A1..A20)
Variance s^2	=VAR(A1:A20)	@vars(A1..A20)

Notice that the letter P, for population (Microsoft Excel), or s, for sample (Lotus 1-2-3), distinguishes the sample-based estimate from the population parameter.

Measure of Population Variability	Microsoft Excel	Lotus 1-2-3
Standard deviation σ	=STDEVP(A1:A20)	@std(A1..A20)
Variance σ^2	=VARP(A1:A20)	@var(A1..A20)

21   If your data are in cells D1 through D15, and you want to display an estimate of the standard deviation in cell E5 and an estimate of the variance in E6, what formula will you enter in cell E5? (Answer the question for the spreadsheet program you use.) _____ In E6? _____

Answer:

If you use Microsoft Excel, E5 would be =STDEV(D1:D15);
E6 would be =VAR(D1:D15).

If you use Lotus 1-2-3, E5 would be @stds(D1..D15);
E6 would be @vars(D1..D15).

22 If your data are in cells D1 through D375, and you want to display the population standard deviation in cell E6, what formula will you enter in cell E6? (Answer the question for the spreadsheet program you use.)

_____

Answer:

If you use Microsoft Excel, =STDEVP(D1:D375).
If you use Lotus 1-2-3, @std(D1..D375).

23 What spreadsheet function corresponds to *s*? _____
What spreadsheet function corresponds to σ? _____
(Answer the question for the spreadsheet program you use.)

Answer:

If you use Microsoft Excel, *s* corresponds to STDEV( ) and σ corresponds to STDEVP( ).
If you use Lotus 1-2-3, *s* corresponds to @stdevs( ) and σ corresponds to @stdev( ).

CONFIDENCE INTERVALS FOR μ

The amount of confidence you can put in an estimate of a parameter varies. In some cases you can say that it is virtually inconceivable for your estimate to be wrong by an amount that has any practical significance. In other cases you must consider your estimate, at best, a "ballpark" figure—an indicator only of the approximate magnitude of the parameter of interest. One useful way to deal with this problem is to establish a confidence interval for your estimate; for example, you might say "the mean number of characters per message in our data communications system is between 307 and 313 with 98% confidence."

24 To clarify how you can establish a confidence interval, and what it means, let us consider a somewhat artificial example. Suppose you are trying to estimate the mean of a population in a situation in which you know the population variability.

A factory has two machines that cut spaghetti and noodles into appropriate lengths for packaging. Both machines can be adjusted to cut any given length. The variability of the two machines differs, however. The standard deviation of pasta cut on machine A is 0.1 cm. The standard deviation of pasta cut on machine B is 1.0 cm.

If the factory manager wants all noodles in a batch to be as nearly as possible the same length, which machine will he use to cut them?

Answer: Machine A, because its output is less variable.

25 Now consider what would happen if you took a number of samples of pasta cut by machine A and computed the mean of each sample. Suppose, for example, that the machine is set to cut pasta 30 cm in length and you take a number of 36-noodle samples. We can predict the sampling distribution for machine A on the basis of the central limit theorem. Complete the following table:

Machine A

Population: $\mu = 30$ cm

$\sigma = 0.1$ cm

Samples of 36: $\mu_{\bar{x}} = \mu = $ _____

$\sigma_{\bar{x}} = \dfrac{\sigma}{\sqrt{n}} = $ _____

Answer:

$$\mu_{\bar{x}} = \mu = 30 \text{ cm}$$

$$\sigma = \frac{0.1}{\sqrt{36}} = \frac{0.1}{6} = 0.0167 \text{ cm}$$

26 According to the normal distribution table, how often will $\bar{x}$ be within 0.0167 cm of μ?

Answer: 0.6826. Roughly 68 percent of the time. Since $\sigma_{\bar{x}} = 0.0167$, $z = 1.0$. The entry in the normal distribution table for $z = 1.0$ is 0.3413. You are interested in values either above or below μ, so you must double the value in the table.

27 "Under the circumstances described (σ = 0.1, n = 36), $\bar{x}$ will be within 0.0167 cm of μ 68 percent of the time." True or false?

Answer: True

28 Now, suppose that m is not known. If σ and n remain unchanged, the above statement will be true no matter what the value of μ. If μ is unknown, there is a 68 percent chance that we will be correct in predicting that it is within 0.0167 cm of whatever $\bar{x}$ we compute from a sample; for example, if we obtain an $\bar{x}$ of 20 from our sample, there is a 68 percent chance that μ is between 20 + 0.0167 and 20 – 0.0167.

We can say that μ is between 19.9833 and _____ with 68% confidence.

Answer: 20.0167

29 If you are going to draw practical conclusions from your estimates, you will probably want a higher confidence level than 68%; for example, you might want to take only a 5% risk of making an incorrect estimate. Then you would choose a confidence level of _____ .

Answer: 95%

30 Assume that you have taken a sample of 36 noodles cut on machine A and have found that the mean length of the 36 noodles is 25 cm. You now want to determine the exact setting of the machine with 95% confidence. Here is how to establish a 95% confidence interval:

(a) Decide what confidence level you will use. We chose 95%.

(b) Divide the confidence level by two to find what percentage of the sampling distribution must be included on each side of the mean. For 95% you must include 47.5% on each side of the mean.

(c) Look in the normal distribution table to find the z score that includes the appropriate percentage. We find that 0.475 corresponds to a z score of 1.96.

(d) The z score tells you how many standard deviations on either side of the mean you must go to include the desired percentage. You must multiply it by the standard deviation of the sampling distribution to obtain a value in centimeters.

$1.96 \times 0.0167 = .033$ cm

(e) Add the value in centimeters to the sample mean to obtain the upper limit of the confidence interval and subtract it from the sample mean to obtain the lower limit. Since the sample mean was 25.000, the upper limit is _____ . The lower limit is _____ .

Answer:

25.033
24.967

31 Now assume that you want to compute a 99% confidence interval on the basis of the same sample:

$$x = 25.000 \text{ cm}$$
$$\sigma = 0.1 \text{ cm}$$
$$n = 36$$
$$\sigma_{\bar{x}} = \frac{\sigma}{\sqrt{n}} = 0.0167$$

Compute a 99% confidence level for μ. Refer to frame 30, if you wish, as you complete these steps.

(a) What is the confidence level?

(b) What percentage of the distribution will be on each side of the mean?

(c) What is the z score?

(d) How many centimeters?

(e) What are the upper and lower limits?

Answer:

(a) 99%

(b) 49.5%

(c) 2.58

(d) 0.043 cm (=2.58 × 0.0167)

(e) 24.957–25.043

32 z_0 is the symbol used to indicate the z score that corresponds to your choice of a confidence level; for example, with a 95% confidence level, $z_0 = 1.96$. With a 99% confidence level, $z_0 =$ _____ .

Answer: 2.58

33 The formula for a confidence interval for μ is

$$\bar{x} \pm z_0 \left(\frac{\sigma}{\sqrt{n}} \right)$$

The symbol ± is read "plus or minus." In this formula, ± tells you to add to find the upper limit and subtract to find the lower limit. Using this formula as a reminder, compute the 99% confidence interval for the setting of machine B on the basis of the following information:

$$\bar{x} = 25$$

$$\sigma = 1.0$$

$$n = 36$$

μ is between _____ and _____ with 99% confidence.

Answer: 24.57 and 25.43

34 What corresponds to a confidence level of 90%?

Answer: 1.65

CONFIDENCE INTERVALS WITH MICROSOFT EXCEL

The Microsoft Excel spreadsheet program provides a function that sets a confidence interval based on the normal distribution. You must give the spreadsheet the percentage of error you are willing to accept, the population standard deviation, and the sample size. The CONFIDENCE(error, σ, n) function will return the amount "plus or minus" from the mean that establishes the confidence interval.

For example, suppose:

You are willing to accept 5% error (95% confidence)

The sample size is 30, cell A5

The population standard deviation is 5, cell A6

The sample mean is 25

CONFIDENCE(0.05, A6, A5) will tell you the ± confidence interval, which you may add to the mean to find the upper limit and subtract from the mean to find the lower limit.

35 Suppose:

You are willing to accept 2% error (98% confidence)

The sample size is 21, cell C2

The population standard deviation is 3.5, cell B3

The sample mean is 17.4, cell B2

What formula will you use to find the upper limit of your confidence interval? _____

Answer: =B2+CONFIDENCE(0.02, A6, A5)

CONFIDENCE INTERVALS FOR μ WHEN σ IS UNKNOWN

When you do not know the population standard deviation, you can still calculate a confidence interval for μ on the basis of sample statistics. To do this, you must know both the mean and the standard deviation of the sample.

36 In the cases we have discussed so far, you are estimating μ but you know the exact value of the population parameter σ. Do you think this would be a common state of affairs?

Answer: No. You are most likely to encounter situations in which both μ and σ are unknown.

37 If the sample size is thirty or more, you can obtain a close estimate of the true confidence interval by substituting an estimate for σ. What is the estimate for σ?

Answer: s

38 Reference formulas:

$$\text{Confidence interval for } \mu: \bar{x} \pm z_0 \frac{\sigma}{\sqrt{n}}$$

$$\text{Estimate of } \sigma: s = \sqrt{\frac{\sum (x - \bar{x})^2}{n - 1}}$$

You are studying the time it takes for an air-traffic controller to read a complex radar display and react to it under certain standardized conditions. By conducting one trial with each of 36 randomly selected controllers, you obtain the following statistics.

$$\bar{x} = 3.15 \text{ seconds; } \sum (x - \bar{x})^2 = 20.00; n = 36$$

What is the 95% confidence interval for mean time to read and react to the display?

Answer:

(a) 95% confidence interval

(b) 47.5% on each side of the mean

(c) $z_0 = 1.96$

(d) $s = \sqrt{\dfrac{\sum (x - \bar{x})^2}{n - 1}} = \sqrt{\dfrac{20}{35}} = 0.755; z_0 \dfrac{s}{\sqrt{n}} = 1.96 \left(\dfrac{0.755}{6} \right) = 0.25$

(e) The confidence interval is 3.15 ±0.25.

μ is between 2.90 and 3.40 with 95% confidence.

39 The normal probability table gives the sampling distribution of x for samples of thirty or more. Can you use it to establish a confidence interval for μ on the basis of a sample of ten?

Answer: No

40 To use the normal probability table for establishing a confidence interval for μ, you must have a sample size of at least _____ .

Answer: 30

41 If your sample is smaller than thirty, it is still possible to establish a confidence interval for μ, provided you can assume the population distribution is normal; for example, suppose in your study of air-traffic controllers you are able to observe only samples of twenty. The graphs below summarize the data you obtain for three different tasks.

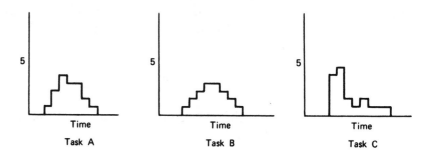

FIGURE 3-1:
Tasks

You are prepared from experience to assume that the population distributions are normal unless the data make this assumption unlikely. Would these data make you assume that the population distribution for any of these tasks is not normal? If so, which one(s)?

Answer: Task C. The sample distribution is so skewed that it seems unlikely to have come from a normally distributed population. (Although the sample distribution for Task A is also slightly skewed, you are quite likely to draw a slightly skewed sample of twenty from a normally distributed population.)

42 For samples larger than thirty, must you assume that the population is normally distributed in order to establish a confidence interval for μ?

Answer: No

43 For samples smaller than thirty, must you assume that the population is normally distributed in order to establish a confidence interval for μ?

Answer: Yes

44 Can you establish a confidence interval for μ on the basis of a sample of fifteen from a population that is not normally distributed?

Answer: No

45 When you establish a confidence interval for μ for a small, normally distributed sample, you must use a sampling distribution called the "Student's t." This distribution is similar to the normal distribution, but its exact shape depends on the size of the sample. The procedure is the same except that

(a) Instead of using a value of z_0 from the normal probability table, you must use a value of t_0 from a t table.

(b) You must always use s, even if σ is known.

For example, let us establish a 99% confidence interval for the mean time of task B in frame 41.

$$\bar{x} = 3.00 \qquad s = 0.53 \qquad n = 20$$

From the t distribution table we find that t_0 for 99% confidence of a sample of 20 is 2.86. (You will learn how to read the table in a moment.)

The formula for the confidence interval of a small, normally distributed sample is

$$\bar{x} \pm t_0 \frac{s}{\sqrt{n}}$$

What is the confidence interval in this case?

Answer: $\bar{x} \pm t_0(s/\sqrt{n}) = 3.00 \pm 2.86 = (0.53/\sqrt{20}) = 3.00 \pm 0.34$; μ is between 2.66 and 3.34 with 99% confidence.

The first column lists the number of degrees of freedom (df). The headings of the other columns give probabilities (P) for t to exceed the entry value. Use symmetry for negative t values.

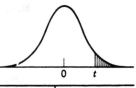

P df	.10	.05	.025	.01	.005
1	3.078	6.314	12.706	31.821	63.657
2	1.886	2.920	4.303	6.965	9.925
3	1.638	2.353	3.182	4.541	5.841
4	1.533	2.132	2.776	3.747	4.604
5	1.476	2.015	2.571	3.365	4.032
6	1.440	1.943	2.447	3.143	3.707
7	1.415	1.895	2.365	2.998	3.499
8	1.397	1.860	2.306	2.896	3.355
9	1.383	1.833	2.262	2.821	3.250
10	1.372	1.812	2.228	2.764	3.169

FIGURE 3-2:

Critical points of the t distribution

46 The above is an excerpt from a t table. Each line represents the distribution of t for a particular sample size. The df (degrees of freedom) column on the left allows you to select the correct distribution for your sample size. The degrees of freedom in a single sample are equal to $n - 1$. Thus the degrees of freedom for a sample of eight are _____ .

Answer: Seven

47 The degrees of freedom for a sample of ten are _____ .

Answer: Nine

48 The headings at the top of the t table give the probabilities that the mean is outside the upper or lower limit set by the indicated values of t. For example, if you are setting a 95% confidence interval, you will want the probability of a mean outside the upper limit to be only 2.5%. (Of course, the probability of a mean outside the lower limit will also be 2.5%.) You will therefore choose a value of t from the column headed 0.025. What column will you use if you are setting a 90% confidence interval?

Answer: The column headed 0.05

49 You want to look up t_0 for a 98% confidence interval for a sample of six. What value of df will you use?

Answer: 5 (df = *n* – 1)

50 What column will you use?

Answer: 0.01

51 Circle the correct entry in the *t* table above frame 46.

Answer:

The first column lists the number of degrees of freedom (df). The headings of the other columns give probabilities (*P*) for *t* to exceed the entry value. Use symmetry for negative *t* values.

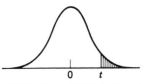

P df	.10	.05	.025	.01	.005
1	3.078	6.314	12.706	31.821	63.657
2	1.886	2.920	4.303	6.965	9.925
3	1.638	2.353	3.182	4.541	5.841
4	1.533	2.132	2.776	3.747	4.604
5	1.476	2.015	2.571	3.365	4.032
6	1.440	1.943	2.447	3.143	3.707
7	1.415	1.895	2.365	2.998	3.499
8	1.397	1.860	2.306	2.896	3.355
9	1.383	1.833	2.262	2.821	3.250
10	1.372	1.812	2.228	2.764	3.169

FIGURE 3-3:
Critical points of
the *t* distribution

52 You want to set a 99% confidence interval for μ on the basis of a sample of fifteen. Look up the appropriate t_0 in table V in the back of the book.

$t_0 = $ _____

Answer: 2.977

You will notice that the t table has entries for df of more than thirty. If the population is normally distributed but σ is unknown, the t table can be used to obtain a precise confidence interval for μ, more accurate than the approximation obtained when you use s as an estimate of σ with the normal distribution table. The larger the sample, the closer t_0 comes to z_0, so that for large samples there is little practical difference between the two procedures.

53 For each of the following samples, how would you go about establishing a 98% confidence interval for μ? Choose the correct answer for each sample.

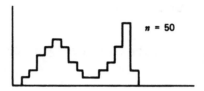

FIGURE 3-4:
Sample A

The population from which this sample was drawn has a known standard deviation: σ = 7.85. The population distribution is known not to be normal.

(a) Use z.

(b) Use t.

(c) Can't establish a confidence interval.

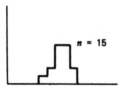

FIGURE 3-5:
Sample B

The population is assumed to have a normal distribution.

(a) Use z.

(b) Use t.

(c) Can't establish a confidence interval.

FIGURE 3-6:
Sample C

The standard deviation of the population is unknown. The distribution is known not to be normal.

(a) Use z.

(b) Use t.

(c) Can't establish a confidence interval.

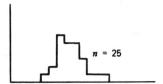

FIGURE 3-7:
Sample D

The population is known to be normally distributed, with a standard deviation: $\sigma = 8.0$.

(a) Use z.

(b) Use t.

(c) Can't establish a confidence interval.

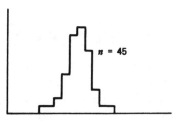

FIGURE 3-8:
Sample E

The population is known to be normally distributed.

(a) Use z.

(b) Use t.

(c) Can't establish a confidence interval.

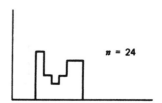

FIGURE 3-9:
Sample F

On the basis of experience you assume the population is not normally distributed.

(a) Use z.

(b) Use t.

(c) Can't establish a confidence interval.

Answer:

Sample A	(a)	Use z. The sample is large.
Sample B	(b)	Use t. The sample is small but from a normal population.
Sample C	(a)	Use z. The sample is large.
Sample D	(b)	Use t. The sample is small but from a normal population.
Sample E	(a)	Use z. The sample is large and from a normal population.
		or
	(b)	Use t. The results will be almost identical, whichever method you use. Theoretically, t is more accurate because you do not know the exact value of σ.
Sample F	(c)	Can't establish a confidence interval. The sample is small and the population is not normally distributed.

LARGE SAMPLE CONFIDENCE INTERVALS FOR *P*

Just as you can find confidence intervals for μ by using the sampling distribution of $\bar{x}$, you can also find confidence intervals for P by using the sampling distributions for p. For example, if half a random sample of 30 visitors at a given resort are men ($p = 0.50$), you can state with 99% confidence that between 0.27 and 0.73 percent of all visitors are male. For small samples, special tables based on the binomial probability distributions are available. For large samples, the binomial probability distribution becomes approximately normal in shape. Therefore, the normal probability table can be used to set a confidence interval for P if the sample is large.

54 Suppose you are trying to estimate the effectiveness of a vaccine. You administer the vaccine to 200 people and then test to determine whether they have acquired an immunity to the disease in question. You find that 185 of 200 have acquired an immunity; that is, $p =$ _____ .

Answer: 0.925 (=185/200)

55 You would like to establish a 99% confidence interval for P in the above test. First you must make sure the sample is large enough. To establish a confidence interval for P, the smaller group in your sample must have at least 10 cases. The smaller group in this case consists of the 15 cases who did not acquire an immunity. Is the sample large enough?

Answer: Yes

56 The formula for establishing a large-sample confidence interval for P is

$$p \pm z_0 \frac{\sqrt{pq}}{\sqrt{n}}$$

In this formula, p, as you know, is the probability of whatever outcome you are concerned with. The probability of any other outcome is q; for example, if p is the probability of acquiring immunity, q is the probability of not acquiring immunity: $p + q = 1$. If $p = 0.925$, $q =$ _____ .

Answer: 0.075

57 Use the formula to establish a confidence interval for p. The z_0 you use is the same z_0 you would use to establish a large-sample confidence interval for μ; P is between _____ and _____ with 99% confidence.

Answer: 0.973 and 0.877

$$p \pm z_0 \frac{\sqrt{pq}}{\sqrt{n}}$$

$$0.925 \pm 2.58 \frac{\sqrt{(0.925)(0.075)}}{\sqrt{200}} = 0.925 \pm 0.048$$

58 In a sample of 100 cases, 20 people who receive a fund-raising letter contribute and 80 do not. Predict the proportion P you would expect to contribute if the same letter were sent to the entire population of 10,000 from which the sample was selected. Use a 95% confidence interval.

Answer:

$$p \pm z_0 \frac{\sqrt{pq}}{\sqrt{n}} = 0.20 \pm 1.96 \frac{\sqrt{(0.20)(0.80)}}{\sqrt{100}} = 0.20 \pm 0.08$$

USING A SPREADSHEET TO LOOK UP
CRITICAL VALUES OF z AND t

To calculate confidence intervals, you select a probability value and then use a table to look up the value of z or t that corresponds to the confidence level you have selected. Spreadsheet programs provide functions that return the value of z or t that corresponds to a given probability level. Using a given probability to arrive at a corresponding z or t score is called an *inverse* function.

Spreadsheets organize the probability information for normal distributions somewhat differently than the tables we have been using. Spreadsheets analyze normal distribution probabilities cumulatively, starting from the left end of the distribution. Look at figure 3-10.

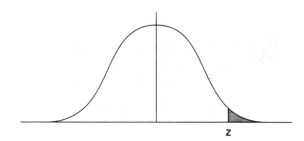

FIGURE 3-10:

Areas under the normal curve

The shaded area under the curve is the probability returned by the spreadsheet functions. As you see, this probability represents the chance that your estimate is wrong in one direction (for example, too high). There is an equal chance that it may be wrong in the other direction (too low). So the probability that you enter into the spreadsheet normal distribution function should be half the possibility of error that you are willing to accept. If you want a 95% confidence interval, you are willing to accept a 5% chance of error. The probability value you enter into the spreadsheet normal distribution function will be 5%/2 = 2.5%, or 0.025.

59 What probability value will you enter into the spreadsheet normal distribution formula if you want a 98% confidence interval? _____

Answer: 2%/2 = 1.0%, or 0.01

60 What probability value will you enter into the spreadsheet normal distribution formula if you want a 99% confidence interval? _____

Answer: 1%/2 = 0.5%, or 0.005

You are working with the standard normal distribution that has $\mu = 0$ and $\sigma = 1.0$. You will need to enter these values so that the value returned by the function is a z score. For Lotus 1-2-3, you also need to identify the function as "type 1" to indicate that you want an inverse function.

	Microsoft Excel	Lotus 1-2-3
Model	=NORMINV(conf, mean, sd)	@normal(conf, mean, sd, type1)
z_0 for 95% confidence (5% chance of error)	=NORMINV(0.025, 0, 1)	@normal(0.025, 0, 1, 1)
z_0 for 99% confidence (1% chance of error)	=NORMINV(0.005, 0, 1)	@normal(0.005, 0, 1, 1)

61 Write the formula you would enter into a cell of your spreadsheet to have your program calculate the z score that corresponds to a 97% confidence interval. (Answer the question for the spreadsheet program you use.) _____

Answer:

If you use Microsoft Excel, =NORMINV(0.015, 0, 1)
If you use Lotus 1-2-3, @normal(0.015, 0, 1, 1)

The spreadsheet functions to look up a critical value of *t* are somewhat different from the normal distribution functions. Spreadsheets organize the probability information for the *t* distribution in "two-tailed" form. Look at figure 3-11.

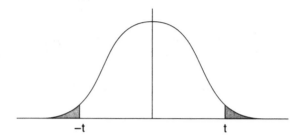

FIGURE 3-11:

The *t* distribution

The probability that you enter into the spreadsheet *t* distribution function should be exactly the same as the possibility of error that you are willing to accept. If you want a 95% confidence interval, you are willing to accept a 5% chance of error. The probability value you enter into the spreadsheet *t* distribution function will be 5%, or 0.05.

62 What probability value will you enter into the spreadsheet *t* distribution formula if you want a 98% confidence interval? _____

Answer: 2%, or 0.02

63 What probability value will you enter into the spreadsheet *t* distribution formula if you want a 99% confidence interval? _____

Answer: 1%, or 0.01

When you use the *t* distribution, you will need to enter the desired probability value and degrees of freedom. For Lotus 1-2-3, you also need to identify the function as "type 1" to indicate that you want an inverse function.

	Microsoft Excel	**Lotus 1-2-3**
Model	=TINV(conf, df)	@tdist(conf, df, type1, tails)
t_0 for 95% confidence, df = 9 (5% chance of error)	=TINV(0.05, 9)	@tdist(0.05, 9, 1, 2)
t_0 for 99% confidence, df = 9 (1% chance of error)	=TINV(0.01, 9)	@tdist(0.01, 9, 1, 2)

 Write the formula you would enter into a cell of your spreadsheet to have your program calculate the t score that corresponds to a 97% confidence interval for a sample size of eleven. (Answer the question for the spreadsheet program you use.) ＿＿＿＿

Answer:

If you use Microsoft Excel, =TINV(0.03, 10)
If you use Lotus 1-2-3, @tdist(0.03, 10, 1, 2)

SELF-TEST

If you have successfully completed this chapter, you can now use sample statistics to estimate population parameters. You can:

- Estimate μ, P, and σ

- Establish a confidence interval for μ, using a sample mean and standard deviation

- Establish a confidence interval for P, using data from a large sample

Now try these review problems. Table I at the back of the book lists any formulas you may need for reference.

1. The following data are a sample selected from a population you assume to be normally distributed. What is your best estimate of μ and σ? Establish a 95% confidence interval for μ.

    Data: 3, 4, 5, 6, 6, 7, 7, 7, 8, 8, 8, 9, 9, 10, 11, 12

2. In an appropriately selected sample of 144 members of a fraternal organization, 20 percent are college graduates. Estimate the percentage you would find if you surveyed all 200,000 members. Establish a 99% confidence interval.

3. You select a random sample of 100 chocolate-covered peanuts from the output of a candy factory. Measuring the thickness of the chocolate coating yields the following statistics for this sample:

$\bar{x} = 0.1$ cm $s = 0.01$ cm

Establish a 95% confidence interval for the parameter μ.

4. State briefly the assumptions involved in establishing a confidence interval using z and t.

ANSWERS

To review a problem, study the frames indicated after the answer.

1. The best estimate of μ is $\bar{x}$.

$$\bar{x} = \frac{\sum x}{n} = \frac{120}{16} = 7.50$$

The best estimate of σ is s.

$$s = \sqrt{\frac{\sum(x - \bar{x})^2}{n - 1}} = \sqrt{\frac{88}{15}} = \sqrt{5.87} = 2.42$$

For a confidence interval you must use t because the sample is small.

$$t_0 = 2.13 \qquad (df = 15)$$

The formula for the confidence interval is

$$\bar{x} \pm t_0 = \frac{s}{\sqrt{n}} = 7.5 \pm 2.13 \frac{2.42}{\sqrt{16}}$$

$$= 7.5 \pm 2.13(0.605) = 7.5 \pm 1.29$$

μ is between 6.21 and 8.79 with 95% confidence. *See frames 36 to 53.*

2. This is a large-sample confidence interval for P.

    The formula is

$$p \pm z_0 \frac{\sqrt{pq}}{\sqrt{n}} = 0.20 \pm 2.58 \frac{\sqrt{(0.20)(0.80)}}{\sqrt{144}}$$

$$= 0.20 \pm 2.58 \frac{0.4}{12} = 0.20 \pm 0.085$$

P is between 0.115 and 0.285 with 99% confidence. *See frames 54 to 58.*

3. Because of the large sample size, the appropriate formula is

$$\bar{x} \pm z_0 \frac{s}{\sqrt{n}} = 0.1 \pm 1.96 \frac{0.01}{\sqrt{100}} = 0.1 \pm 0.00196$$

The mean thickness of the chocolate in the factory's output is between 0.098 and 0.102 cm with 95% confidence. *See frames 24 to 34.*

4. Using *z* to establish a confidence interval assumes that the sample size is more than thirty. Either σ is known or *s* is used as an estimate. Using *t* to establish a confidence interval assumes that the population is normally distributed; σ is never used, always *s*. *See frames 30 to 34, 45 to 48, and 53.*

4 Hypothesis Testing

In scientific research, we develop theories and then conduct experiments to test them. Since we cannot achieve perfect control in any experiment, some chance variation always results. For example, even the most precise measurement system has some limit to its accuracy, and the laboratory mice bred for biological research to be as similar as possible are still not completely identical. Also, there is a limit to the number of times observations can be repeated, so our measurements are always only a sample of reality.

To generalize from our experimental sample, we use statistical techniques to test how big a part chance plays in the outcome of an experiment. We begin by assuming that our experimental results reflect only the random variations caused by assorted factors beyond our control. This assumption is called the *null hypothesis*. If our experiment is successful and our theory is true, we will be able to reject the null hypothesis by showing that chance variation is not a reasonable explanation for our results.

When you have completed this chapter, you will be able to:

- State a null hypothesis and alternative and establish a critical region for theories about P

- Perform a similar statistical test for theories about μ

- Judge the probability of accepting a theory when it is false or rejecting a theory when it is true

HYPOTHESIS TESTING-PROPORTIONS

To understand how statistical tests are used in testing scientific theories, we will begin by looking at how to use the binomial probability table to test theories about proportions.

1 There is a formal procedure for the statistical testing of theories in scientific research:

(a) Plan an experiment so that if the results cannot be explained by the chance variation involved in drawing a sample, your theory will be confirmed.

(b) Conduct the experiment and collect sample data.

(c) Assume the results are due to chance alone. This assumption is called the null hypothesis.

(d) Use a theoretical sampling distribution based on the null hypothesis to determine the probability of obtaining sample data like yours by chance alone.

(e) If the probability of obtaining sample data like yours by chance alone is less than some predetermined small percentage (usually 5% or 1%), the results will be significant. You may reject the null hypothesis and consider your theory confirmed.

In this procedure the null hypothesis assumes that the results of the experiment are due to _____ .

Answer: Chance, the chance variation involved in drawing a sample

2 If you reject the null hypothesis, your theory (is/is not) confirmed.

Answer: Is

3 Study these definitions:

Null hypothesis	Assumption that experimental results are due to chance alone
Alternative	Your theory (will be confirmed if you reject the null hypothesis)

Significant results	Experimental results that are not likely to have occurred by chance alone

The assumption that experimental results are due to chance is called the _____ .

Answer: Null hypothesis

4 Your theory is called the _____ .

Answer: Alternative

5 If the results are significant, you will reject the _____ .

Answer: Null hypothesis

6 Nonsignificant results (will/will not) allow you to reject the null hypothesis.

Answer: Will not

7 Results that are unlikely to have occurred by chance are called _____ .

Answer: Significant

8 Let us look at an example of this procedure. A researcher is studying the behavior of fruit flies. He wants to investigate the theory that "you can catch more flies with honey than with vinegar." He establishes a standardized procedure for using both substances as bait, and he plans to count the total number of flies caught, n, and the proportion caught with honey, p. His null hypothesis is that there is no real difference between the number of flies that can be caught with honey or vinegar. The differences are due only to chance variations from sample to sample. If this is true, the population parameter $P =$ _____ .

Answer: 0.5. Flies show no real preference in how they get caught. They might as well just "flip a coin."

9 The researcher decides to continue the experiment until he catches 15 flies. He can now find a theoretical sampling distribution based on the null hypothesis by looking at a binomial probability table. For $P = 0.5$ and $n = 15$ the distribution is as follows:

14	0.05	0.1	0.2	0.25	0.3	0.4	0.5	0.6	0.7	0.75	0.8	0.9	0.95
15 0	0.463	0.206	0.035	0.013	0.005								
1	0.366	0.343	0.132	0.067	0.031	0.005							
2	0.135	0.267	0.231	0.156	0.092	0.022	0.003						
3	0.031	0.129	0.250	0.225	0.170	0.063	0.014	0.002					
4	0.005	0.043	0.188	0.225	0.219	0.127	0.042	0.007	0.001				
5	0.001	0.010	0.103	0.165	0.206	0.186	0.092	0.024	0.003	0.001			
6		0.002	0.043	0.092	0.147	0.207	0.153	0.061	0.012	0.001	0.001		
7		0.014	0.039	0.081	0.177	0.196	0.118	0.035	0.013	0.003			
8		0.003	0.013	0.035	0.118	0.196	0.177	0.081	0.039	0.014			
9		0.001	0.003	0.012	0.061	0.153	0.207	0.147	0.092	0.043	0.002		
10			0.001	0.003	0.024	0.092	0.186	0.206	0.165	0.103	0.010	0.001	
11				0.001	0.007	0.042	0.127	0.219	0.225	0.188	0.043	0.005	
12					0.002	0.014	0.063	0.170	0.225	0.250	0.129	0.031	
13						0.003	0.022	0.092	0.156	0.231	0.267	0.135	
14							0.005	0.031	0.067	0.132	0.343	0.366	
15								0.005	0.013	0.035	0.206	0.463	

FIGURE 4-1:
Binomial
probabilities

How often would he expect to catch 10 or more flies with honey by chance alone?

Answer: 0.151, or 15.1% of the time. You must add the frequencies for 10, 11, 12, . . . , 15. Note: the blank areas in the table represent probabilities so small that they round off to zero.

10 The researcher decides that he will reject the null hypothesis only if his results would occur by chance less than 2% of the time. How many flies will he have to catch with honey to reject the null hypothesis? Use figure 4-1.

Answer: At least 12. The frequencies for 12, 13, 14, and 15 add up to 0.017, or 1.7%. Catching 11 or more flies would occur by chance 0.059, or 5.9% of the time, so 12 is the smallest number that will meet the researcher's criterion.

11 The researcher catches thirteen flies with honey and two with vinegar. Are his results significant?

Answer: Yes

12 The set of results that are significant is called the critical region. In this case the critical region is $p \geq 12/15$ (p is equal to or greater than $12/15$). Is $14/15$ in the critical region?

Answer: Yes

13 Is 12/15 in the critical region?

Answer: Yes

14 For this experiment, $p \geq 12/15$ is called the _____ .

Answer: Critical region

15

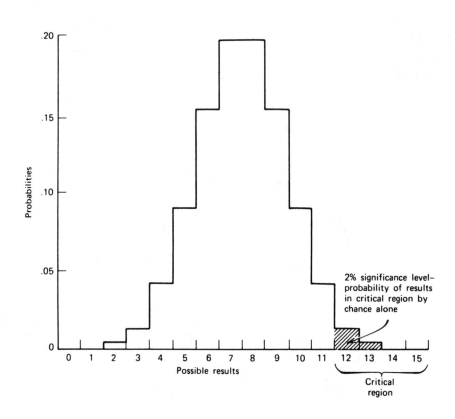

FIGURE 4-2:

Probability distribution for fruit fly experiment

The significance level is the probability that significant results will occur by chance. For this experiment the researcher is using a significance level of _____ %.

Answer: 2%

16 The probability of obtaining results in the critical region by chance is called the _____ .

Answer: Significance level

17 We can summarize the statistical test of the researcher's hypothesis as follows:

Null hypothesis	$P = 0.5$
Alternative	$P > 0.5$ (P is greater than 0.5)
Significance level	2%
Critical region	$p \geq 12/15$ (p is equal to or greater than 12/15)

If, in fact, more flies can be caught with vinegar than with honey, will the researcher find it out from this statistical test?

Answer: No. Of course, he could have started out with the theory that you can catch more flies with vinegar, considered a fly caught with vinegar a success, and used a similar test.

18 Let us restate the researcher's hypothesis slightly and see how this affects the statistical test. "You will catch a different number of flies with honey than with vinegar." If you can catch flies only with vinegar ($P = 0$), will this theory be confirmed?

Answer: Yes

19 If you can catch flies only with honey ($P = 1$), will this theory be confirmed?

Answer: Yes

20 What is the appropriate null hypothesis to test this theory?

Answer: $P = 0.5$

21   Which of the following is the appropriate alternative for a statistical test of this theory?

(a) $P > 0.5$ (P is greater than 0.5)

(b) $P < 0.5$ (P is less than 0.5)

(c) $P \neq 0.5$ (P is not equal to 0.5)

Answer:

(c) $P \neq 0.5$

22   Which of the illustrations below shows an appropriate critical region for this statistical test?

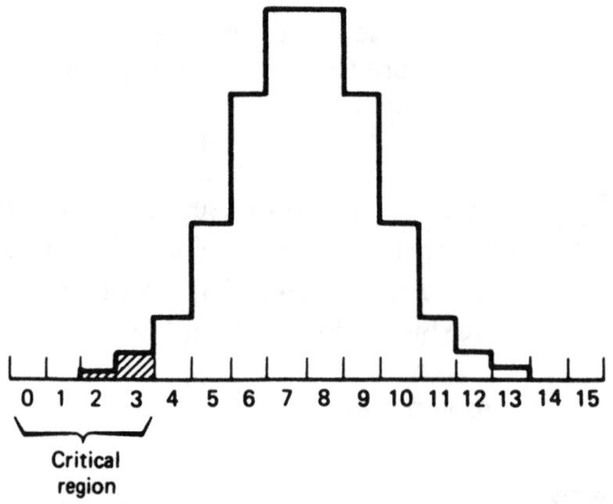

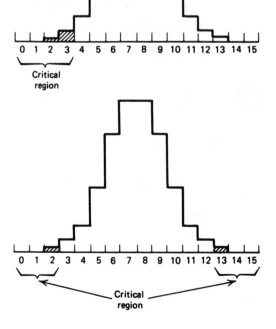

FIGURE 4-3a:
One-tailed probability distribution

FIGURE 4-3b:
Two-tailed probability distribution

Answer:

(b) In this case either a very high p or a very low p would be consistent with the alternative and would cause you to reject the null hypothesis.

23 To establish the critical region for this statistical test, you must consider both ends of the sampling distribution; for example, how often would you expect $p \le 3/15$ (p equal to or less than $3/15$) or $p \ge 12/15$ (p equal to or greater than $12/15$)?

Answer: 0.034, or 3.4 percent of the time. You must add the frequencies for $0, \ldots, 3$ and $12, \ldots, 15$.

24 If the researcher continues to use a significance level of 2%, what will his critical region be?

p_____ or p_____

Answer:

$$p \le 2/15 \text{ or } p \ge 13/15$$

25 Summarize the statistical test for the theory that "you will catch a different number of flies with honey than with vinegar."

Null hypothesis	_____
Alternative	_____
Significance level	2%
Critical region	_____

Answer:

Null hypothesis	$P = 0.5$
Alternative	$P \ne 0.5$
Critical region	$p \le 2/15 \text{ or } p \ge 13/15$

26 A caution before we go on. The statistical test of a hypothesis takes into account the chance variations involved in drawing a sample. It does not take into account unplanned influences in the way we conduct the experiment. In a poorly planned experiment we may obtain statistically significant results that are meaningless; for example, suppose that there is a prevailing wind in the direction of the honey at the time we conduct our fruit fly experiments. We catch $13/15$ of the flies with honey—a statistically significant result. Does this result support the theory that you can catch more flies with honey than with vinegar? Why or why not?

Answer: No. You cannot tell whether it was the honey or the direction of the wind that determined the outcome of the experiment.

27 You are studying gender-related issues in the way children are raised. You want to test the theory that pet ownership is related to gender. In particular, you want to test the theory that one sex will be more heavily represented than the other among owners of frogs. In an appropriately selected sample of fifteen frog owners, you count the proportion of girls, p. What is an appropriate null hypothesis for a statistical test of your theory?

Answer: $P = 0.5$

28 What is an appropriate alternative?

Answer: $P \neq 0.5$. You have no firm prediction that girls or boys are more likely to own frogs.

29 Use figure 4-1 in frame 9 to determine a critical region for a significance level of 5%. What is the critical region?

Answer:

$$p \leq 3/15 \text{ or } p \geq 12/15$$

30 You want to test the theory that more girls than boys own frogs. In an appropriately selected sample of fifteen frog owners, you count the proportion of girls, p. Summarize an appropriate statistical test for your theory, using a 5% significance level.

Null hypothesis   _____

Alternative   _____

Critical region   _____

Answer:

Null hypothesis	$P = 0.5$
Alternative	$P < 0.5$
Critical region	$p \leq 3/15$

You want to test the theory that more girls than boys own frogs. Your sample consists of all twenty-five entrants in the Maricopa County frog-jumping contest sponsored by the local chapter of the Girl Scouts. The results are statistically significant. Comment on the meaning of these results.

Answer: The experiment was poorly planned. All frog owners are not assured an equal probability of being selected for the sample. As a result, the data are meaningless as a test of your theory.

USING A SPREADSHEET TO LOOK UP CRITICAL VALUES OF THE BINOMIAL DISTRIBUTION

As you may suspect, adding up the frequencies for various possible numbers of successes in a binomial distribution is the kind of thing a spreadsheet program can do for you. Just as you can use a spreadsheet to look up critical values of z and t, you can also look up critical values of the binomial distribution. The equivalent of NORMINV( ) for the binomial distribution is CRITBINOM( ). The "critical binomial" functions return the number of successes that corresponds to a given cumulative sampling probability level.

The critical binomial functions require you to provide the number of trials (n), the null hypothesis probability (P) and the cumulative sampling probability (p). They return the number of successes required to at least equal the cumulative sampling probability (p).

	Microsoft Excel	Lotus 1-2-3
Model	=CRITBINOM(n, P, p)	@critbinomial(n, P, p)
15 trials, $P = 0.05$ 5% cumulative p	= CRITBINOM(15, 0.5, 0.05)	@critbinomial(15, 0.5, 0.05)
15 trials, $P = 0.05$ 98% cumulative p	= CRITBINOM(15, 0.5, 0.98)	@critbinomial(15, 0.5, 0.98)

Spreadsheets analyze binomial distribution probabilities cumulatively, starting from the left end of the distribution, in the same way as they treat normal distribution probabilities. See figure 4-4a, b.

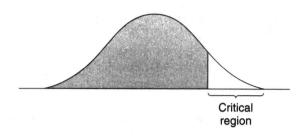

FIGURE 4-4a:
Binomial probability; unshaded area represents critical region

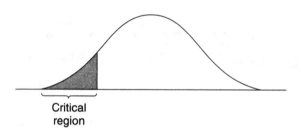

FIGURE 4-4b:
Binomial probability; shaded area represents critical region

The shaded area under the curve is the cumulative sampling probability you will enter into the spreadsheet function. The number the function returns represents the number of successes that it takes to add up to the cumulative sampling probability.

You need to think about your experiment in relation to this shaded area in order to select the appropriate cumulative sampling probability to use. Sometimes the unshaded area (figure 4-4a) will represent the critical region and sometimes the shaded area will represent the critical region (figure 4-4b).

- *Unshaded area represents the critical region.* If a high number of successes represents the situation that would allow you to reject the null hypothesis (figure 4-4a), the cumulative sampling probability you enter in the formula should equal 1.00 minus the significance level (for example, $1.00 - 0.02 = 0.98$). The number the function will return is the highest number that is not yet within the critical region. For example, if the function returns 12, the critical region is $s \geq 13$.

- *Shaded area represents the critical region.* If a low number of successes represents the situation that would allow you to reject the null hypothesis (figure 4-4b), the cumulative sampling probability you enter in the formula should equal the significance level you have selected for your experiment. The number the function will return is the lowest number that is no longer within the critical region. For example, if the function returns 3, the critical region is $s \leq 2$.

- *Two-tailed test.* If your hypothesis requires a two-tailed test, treat the situation as two separate tests, one for half the significance level at the low number end of the sampling distribution and one for the other half at the high end of the sampling distribution.

32 Let's apply the spreadsheet formula to the fruit fly example. Our fruit fly specialist is going to test the hypothesis that "you can catch more flies with honey than with vinegar" in a series of fifteen trials. He has decided on a significance level of 2%. The critical region is in the (shaded/unshaded) area. _____

Answer: The unshaded area. A high number of successes will allow him to reject the null hypothesis.

33 Write the formula you would enter into a cell of your spreadsheet to have your program calculate the critical region for this experiment. (Answer the question for the spreadsheet program you use.) _____

Answer:

If you use Microsoft Excel, = CRITBINOM(15, 0.5, 0.98)
If you use Lotus 1-2-3, @critbinomial(15, 0.5, 0.98)

34 The spreadsheet function will return a value of 11. This tells you that the critical region for the test is _____ .

Answer:

$$p \geq 12/15$$

35 For additional practice, let us revisit the question of who owns frogs. You have no firm prediction whether girls or boys are more likely to own frogs, but you expect that frog owners are not equally likely to be of either gender. You set up an appropriate test of your hypothesis for a random group of twenty-five frog owners, using a 5% significance level.

Null hypothesis	$P = 0.5$
Alternative	$P \neq 0.5$
Significance level	5%

What spreadsheet formulas will tell you the critical region for this experiment? (Answer the question for the spreadsheet program you use.)

_____

Answer:

You must find a 2.5% critical region at each end of the sampling probability distribution.
If you use Microsoft Excel, = CRITBINOM(25, 0.5, 0.975) and =CRITBINOM(25, 0.5, 0.025)
If you use Lotus 1-2-3, @critbinomial(15, 0.5, 0.98) and @critbinomial(25, 0.5, 0.025)

36   If you use the formulas, you will find that they return the values 17 and 8. What is the critical region?

Answer: The critical region is $p \leq 7/25$ or $p \geq 18/25$

HYPOTHESIS TESTING—MEANS

So far you have used the hypothesis-testing procedure for theories about the parameter P, but you can use the same kind of procedure for theories about μ.

37   Our fruit fly researcher believes feeding fruit flies a mixture of honey and vinegar will alter their life span. He knows from long experience that the particular strain of fruit fly he is working with has a mean life span of twelve days, with a standard deviation of two, when fed their normal diet of apple juice. He plans to raise a sample of fifty fruit flies on honey and vinegar and calculate their mean life span. What is an appropriate null hypothesis for a statistical test of this theory? _____

Answer: $\mu = 12$

38   What is the alternative?

Answer: $\mu \neq 12$

39 The **Central Limit Theorem** states that for large samples (n > 30),

$$\mu_{\bar{x}} = \mu, \qquad \sigma_{\bar{x}} = \frac{\sigma}{\sqrt{n}}$$

The central limit theorem allows the researcher to develop a sampling distribution based on the null hypothesis. The mean of the sampling distribution for the fruit fly experiment will be _____ .

Answer: 12

40 The standard deviation of the sampling distribution will be _____ = 0.28.

Answer: $\frac{2}{\sqrt{50}}$

41 The procedure for establishing a critical region in this case is the same as that for establishing a confidence interval for μ. If our fruit fly researcher uses a 1% significance level (equivalent to a 99% confidence interval), what will be the critical region? $z \geq$ _____ or $z \leq$ _____ .

Answer:

$$z \geq 2.58 \text{ or } z \leq -2.58$$

42

$$z_{\bar{x}} = \frac{\bar{x} - \mu_{\bar{x}}}{\sigma_{\bar{x}}}, \text{ or } \frac{\bar{x} - \mu}{(\sigma/\sqrt{n})}$$

If the researcher finds that the mean life span of his sample of 50 fruit flies is 12.5 days, can he reject the null hypothesis?

Answer: No (z = +1.77)

43 Summarize the statistical test:

Null hypothesis _____

Alternative _____

Critical region _____

Answer:

Null hypothesis	$\mu = 12$
Alternative	$\mu \neq 12$
Critical region	$z \geq +2.58$ or $z \leq -2.58$

44 A company has administered a brief verbal-skills test to all employees for a number of years. For all 6,000 present employees, the mean score is 50 with a standard deviation of 10. A personnel researcher wishes to investigate the theory that first-line supervisors score higher on this test than the average employee. To test this theory she selects a random sample of 100 first-line supervisors and computes their mean score on the test. In this problem you are concerned with only one end of the sampling distribution. Summarize an appropriate statistical test to confirm this theory at the 1% level of significance.

Null hypothesis   _____

Alternative   _____

Critical region   _____

Answer:

Null hypothesis	$\mu = 50$
Alternative	$\mu > 50$
Critical region	$z \geq +2.33$

45 If the mean score of the sample is 53, are the results significant? Refer to the reference formulas in table I at the back of the book if you wish.

Answer: Yes

$$z = \frac{\bar{x} - \mu_{\bar{x}}}{\sigma_{\bar{x}}} = \frac{\bar{x} - \mu}{(\sigma/\sqrt{n})} = \frac{53 - 50}{(10/\sqrt{100})} = 3.0$$

46 As you learned in chapter 3, it is possible to use a spreadsheet program rather than a table to look up critical values of z. What formula would you use to find the critical value of z for the experiment described in frame 44? (Answer the question for the spreadsheet program you use. If you need to, refer back to chapter 3.)

Answer:

If you use Microsoft Excel, = NORMINV(0.01, 0, 1)
If you use Lotus 1-2-3, @normal(0.01, 0, 1, 1)

THE PROBABILITY OF ERROR

When you test theories using statistical procedures, there is always a certain possibility that an unusual randomly selected sample will lead you astray. You cannot eliminate the possibility of error, but you can calculate how great the chance of error is and, if necessary, change your experiment to keep it within acceptable limits.

47 A statistical test of a theory never gives absolute certainty. Even when an experiment is planned perfectly, there are two possible types of error:

Type I   The theory is not true (the null hypothesis is true) but the results are significant by chance.

Type II   The theory is true (the null hypothesis is false) but the results are not significant.

As an example, if you use a 5% significance level, you will obtain significant results by chance 5% of the time. This is an example of a Type ___ error.

Answer: Type I error

48 A Type I error occurs when the theory is (true/false) and the results are (significant/not significant).

Answer:

False
Significant

49 If the theory is true but the results are not significant, a Type ___ error occurs.

Answer: II

50 A Type II error occurs when the theory is (true/false) and the results are (significant/not significant).

Answer:

True
Not significant

51 Label the situations that correspond to Type I and Type II errors:

(a) Theory false, results significant

(b) Theory true, results significant

(c) Theory false, results not significant

(d) Theory true, results not significant

Answer:

(a) Theory false, results significant; Type I error

(d) Theory true, results not significant; Type II error

52 In 100 similar studies of the differences between two teaching methods, 75 studies result in significant differences between the two methods and 25 find no significant difference. If you believe that all the studies were well designed, you will probably consider the 25 studies in which no significant difference was found to be examples of Type _____ errors.

Answer: II

53 Seventy-five studies are conducted to determine the effect of a particular drug on the appetite of white rats. The sample sizes vary, and several different measures of appetite are used. Four of the studies have results significant at the 5% level. What type of error do you think may be involved here? _____

Answer: Type I. In a large number of studies you would expect a certain number of significant results by chance alone.

54 The significance level you use determines the probability of a Type _____ error.

Answer: I

55 With a significance level of 1%, the probability of a Type I error is _____.

Answer: 0.01 (1% of the time)

56 The Greek letter alpha (α) is used as the symbol for the probability of a Type I error. For a statistical test using a significance level of 5%, we say α = 0.05. For a statistical test using a significance level of 2%, we say _____ .

Answer: α = 0.02

57 The Greek letter beta (β) is used as the symbol for the probability of a Type II error. Thus the symbol for the probability of Type I error is _____ and the symbol for the probability of Type II error is _____ .

Answer:

α

β

58 β is the probability of a Type _____ error. α is the probability of a Type _____ error.

Answer:

II

I

59 If a researcher's theory is true and β is very large, the chances of proving the theory with a statistical test are (excellent/poor).

Answer: Poor. He is likely to make a Type II error.

60 Instead of referring to β, researchers sometimes use the term "power." Power is (1 − β). A statistical test with a low β has a (high/low) power.

Answer: High

61 A statistical test for which there is a high probability of a Type II error has (high/low) power.

Answer: Low

62 Would you say that power in a statistical test is a good thing?

Answer: Yes. You have a better chance of confirming your theories if they are correct.

63 If you make an assumption about the true state of affairs, you can determine the power of a statistical test; for example, let us consider the test at the start of this chapter for the theory "You can catch more flies with honey than with vinegar." As you recall, the experiment consisted of catching fifteen flies and counting the number caught with honey as opposed to vinegar. A significance level of 2% was chosen ($\alpha = 0.02$).

Null hypothesis $P = 0.5$

Alternative $P > 0.5$

Critical region $P \geq 12/15$

To reject the null hypothesis we must catch at least _____ of the 15 flies with honey.

Answer: 12

64 Let us assume that if we could test all flies, 80 percent would prefer honey ($P = 0.8$). We can now use the binomial probability table to give us a sampling distribution for samples of 15 from this population. The distribution is as shown below.

14

		0.05	0.1	0.2	0.25	0.3	0.4	0.5	0.6	0.7	0.75	0.8	0.9	0.95
15	0	0.463	0.206	0.035	0.013	0.005								
	1	0.366	0.343	0.132	0.067	0.031	0.005							
	2	0.135	0.267	0.231	0.156	0.092	0.022	0.003						
	3	0.031	0.129	0.250	0.225	0.170	0.063	0.014	0.002					
	4	0.005	0.043	0.188	0.225	0.219	0.127	0.042	0.007	0.001				
	5	0.001	0.010	0.103	0.165	0.206	0.186	0.092	0.024	0.003	0.001			
	6		0.002	0.043	0.092	0.147	0.207	0.153	0.061	0.012	0.003	0.001		
	7			0.014	0.039	0.081	0.177	0.196	0.118	0.035	0.013	0.003		
	8			0.003	0.013	0.035	0.118	0.196	0.177	0.081	0.039	0.014		
	9			0.001	0.003	0.012	0.061	0.153	0.207	0.147	0.092	0.043	0.002	
	10				0.001	0.003	0.024	0.092	0.186	0.206	0.165	0.103	0.010	0.001
	11					0.001	0.007	0.042	0.127	0.219	0.225	0.188	0.043	0.005
	12						0.002	0.014	0.063	0.170	0.225	0.250	0.129	0.031
	13							0.003	0.022	0.092	0.156	0.231	0.267	0.135
	14								0.005	0.031	0.067	0.132	0.343	0.366
	15									0.005	0.013	0.035	0.206	0.463

FIGURE 4-5:
Binomial probabilities

What is the probability of catching twelve or more flies with honey from this population?

Answer: 0.648, or about 65% of the time

---

65   We can say that the power of our test against the alternative $P = 0.8$ is 0.648. The chance of rejecting the null hypothesis in this case is 0.648. Therefore the chance of rejecting it incorrectly is _____ .

Answer:

$$1 - 0.648 = 0.352$$

---

66   If $P = 0.8$, how often will this statistical test result in a Type II error? $\beta =$ _____ .

Answer: 0.352, or about 35% of the time

---

67   What is the power of this test against the alternative $P = 0.9$? You will use the same reasoning but now you must use the sampling distribution for $P = 0.9$.

Answer: The power of the test against the alternative $P = 0.9$ is 0.945.

You can make similar computations for statistical tests about the mean and for other statistical tests covered in this book. The details of the computations are not presented, but the logic of the process remains the same as in the simple case we have considered. You must make enough assumptions about the alternative to allow you to construct a theoretical sampling distribution for the alternative; for example, for a test about the mean you must assume values of μ and σ. Then compare this new sampling distribution with the critical region established for your statistical test to determine how often you would obtain significant results under the new assumption.

---

68   Three factors that influence the probability of a Type II error are significance level, size of the sample, and variability of the population. A large α results in a relatively smaller β; for example, an α of 0.05 will cause (more/less) risk of Type II error than an α of 0.01.

Answer: Less. You are more likely to reject the null hypothesis when it is untrue.

69 An α of 0.05 will cause (more/less) risk of a Type I error than an α of 0.01.

Answer: More

70 A large sample will result in relatively less risk of a Type II error; β against a given alternative will be greater with an n of (35 or 100).

Answer: 35

71 If the populations have relatively large standard deviations, β will be relatively greater. In which situation is the risk of a Type II error greater:

(a) σ appears to be about 2.

(b) σ appears to be about 10.

Answer: (b)

72 All other things being equal, the power of a statistical test will be greater with an α of (0.05 or 0.01).

Answer: 0.05

73 All other things being equal, the power of a statistical test will be greater with an n of (30 or 75).

Answer: 75

74 All other things being equal, the power of a statistical test will be greater if the standard deviations of the data appear to be in the range of (5 or 25).

Answer: 5

75 You are screening anticancer drugs for possible beneficial effects. You know that none of them is harmful but you are uncertain whether they do any good. If your results are significant for a given drug, it will be the subject of further intensive investigation. If your results are not significant, the investigation will be dropped. Which type of error would you be more concerned about avoiding (Type I/Type II)?

Answer: Type II. This error would result in discarding a useful drug. A Type I error would result only in wasted effort in further investigation.

76 In this situation you would probably choose an α of (0.05 or 0.001).

Answer: 0.05

77 You are investigating a theory that runs completely counter to all accepted theory in your field. In some preliminary studies with small samples, you obtained results significant at the 5% level. You have now designed an experiment with a very large sample. You have refined your measurement techniques so that you believe the standard deviations of your data will be relatively small. You will probably choose an α of (0.05 or 0.01).

Answer: 0.01. In this case you are most concerned about avoiding a Type I error.

SELF-TEST   If you have successfully completed this chapter, you can now set up a formal statistical test of a theory. You can:

- State a null hypothesis and an alternative
- Establish a critical region for p to test a hypothesis about P
- Establish a critical region for $\bar{x}$ to test a hypothesis about μ
- Comment on the practical importance of Type I and Type II errors in given situations

Now try these review problems. Table I at the back of the book lists any formulas you may need for reference.

1. "We conducted an experiment based on our theory and the results were statistically significant at the 1% level. Therefore our theory is conclusively proved." Comment on this assertion.

2. "For this proposed experiment the probability of a Type II error appears to be so high that it is not worth the effort to collect the data." What does this mean?

3. A clairvoyant claims he can predict whether a coin will land heads or tails. In an experiment that tested his claim, a coin was tossed ten times and he predicted correctly eight times. Outline an appropriate statistical test. Are these results significant at the 5% level?

4. How often would you expect to obtain eight or more correct predictions by chance in the experiment in problem 3?

5. The mean height of adult males in a particular ancient culture was 5 ft. 2 in. with a standard deviation of 2 in., as determined by measurement of a large number of skeletons found in burial sites. A new site has been found that differs somewhat from the others. The discoverer theorizes that the skeletons at this site are from a different genetic group, with a different mean height. Outline an appropriate statistical test with a 1% significance level.

ANSWERS

To review a problem, study the frames indicated after the answer.

1. The statistical test indicates that the results are not likely to have occurred as a result of the chance variations involved in sampling. You would, however, have to examine the design of the experiment carefully to be sure that the experimenter's theory is the only reasonable explanation of the results. Even if the experimental design were perfect, there would be a 1% chance that the theory would not be correct. *See frames 26, 31, and 47 to 77.*

2. Even if the theory were true, you are not likely to obtain results that would allow you to reject the null hypothesis. You are unlikely to obtain significant results. *See frames 47 to 77.*

3. Let us call a correct prediction a success. On the average one should predict correctly half the time just by chance.

Null hypothesis	$P = 0.5$
Alternative	$P > 0.5$
Significance level	$\alpha = 0.05$
Critical region	$p \geq 9/10$

Since $p = 8/10$, the results are not significant. *See frames 1 to 31.*

4. The probabilities in the binomial probability table are

8	0.044
9	0.010
10	0.001
	0.055

You would expect eight or more successes 5.5% of the time. *See chapter 2, frames 52 to 71.*

5.
Null hypothesis	$\mu = 5$ ft 2 in.
Alternative	$\mu \neq 5$ ft 2 in.
Significance level	$\alpha = 0.01$
Critical region	$z \leq -2.58$ or $z \geq +2.58$

See frames 32 to 46.

5 Differences between Means

This chapter is about statistical tests we can use to investigate the difference between the means of two sets of observations. You will learn about two situations. One is the case in which your observations come in pairs. For example, you may have "before" and "after" observations, or you may have observations of situations that are carefully matched except for one factor whose effect you are investigating. The other case is one in which you have two samples selected at random.

In each of these cases, we can use a statistical test to see if the differences can be explained by chance variations or if we may reject this null hypothesis and say that the differences support an experimental hypothesis.

When you have completed this chapter, you will be able to:

- Test the significance of a set of difference scores

- Test the significance of the difference between the means of two independent samples

- Recognize when you have independent samples, and when you have difference scores, and apply the appropriate test procedure to each

DIFFERENCE SCORES

Difference scores are the result of taking two related observations, for example "twin A" and "twin B," or "before" and "after." When dealing with paired data of this sort, it is important to remember that the statistical analysis is based on the differences between the paired scores.

1 One common type of experimental study is the "before" and "after" study. Let us consider an example. We wish to test the theory that people will lose weight if they stick to a diet of grapefruit and whole wheat toast. To test this theory, we select a sample of 49 people, weigh them at the start of our study, and weigh them again after a period of dieting. For each person we subtract the "before" weight from the "after" weight to obtain a difference score; for example, a person who weighs 125 lb at the start of the study and 118 lb at the end of the study will have a difference score of _____ .

Answer: – 7

2 A person who weighs 140 lb at the start of the study and (because he loves whole wheat toast) 150 lb at the end of the study will have a difference score of _____ .

Answer: +10

3 If on the average the diet makes no difference, the mean difference score for the population is _____ .

Answer: Zero

4 If, on the average, people lose weight on this diet, the mean difference score is _____ .

Answer: Less than zero (negative)

5 What is an appropriate null hypothesis for a statistical test of our theory that people will lose weight on this diet?

Answer: The mean difference score for the population is zero.

6 What is an appropriate alternative?

Answer: The mean difference score for the population is less than zero.

7 If we can construct a theoretical sampling distribution based on the null hypothesis, we can make a statistical test of our theory. Let us consider what the population of difference scores would be like if the diet had no effect. We have said that the mean difference score would be zero. Does this mean that the difference score of every person who adheres to the diet will be zero?

Answer: No. Some will lose weight for reasons that have nothing to do with the diet. Others will gain weight. Among a sufficiently large number of dieters, these variations should tend to cancel out to produce a mean of zero.

8 In the preceding problems in which you tested a hypothesis about the mean, your null hypothesis was based on a population whose mean and standard deviation were known. In this case the null hypothesis is based on a theoretical analysis. We can say that if the diet has no effect, the mean of a population of difference scores should be zero, but we have no basis in theory for saying what the standard deviation of this population should be. Instead, we must use the sample to provide us with an estimate of σ. What statistic will we use?

Answer: s

9 As you recall,

$$\sigma_{\bar{x}} = \frac{\sigma}{\sqrt{n}}.$$

Since we do not know σ in this case, our best estimate is $\sigma_{\bar{x}} = $ _____ .

Answer:

$$\frac{s}{\sqrt{n}}$$

10 With this estimate of $\sigma_{\bar{x}}$, we can compute a z score and test our hypothesis.

Null hypothesis $\mu = 0$

Alternative $\mu < 0$

Significance level $\alpha = 0.05$

Critical region $z \leq -1.65$

Our sample provides the following statistics:

$n = 49$

$\bar{x} = -5.0$

$s = 3.5$

$$\text{Compute } z = \frac{\bar{x} - \mu}{\sigma_{\bar{x}}} = \frac{\bar{x} - \mu}{s/\sqrt{n}} = \underline{\hspace{2cm}}.$$

Are the results statistically significant?

Answer:

$$\text{Compute } z = \frac{-5.0 - 0}{3.5/\sqrt{49}} = \frac{-5.0}{3.5/7} = \frac{-5.0}{0.5} = -10.0$$

The results are significant.

11 In this example we used s as an estimate of σ. This estimate is good only if the sample size is _____ .

Answer: Larger than 30

12 Let's review this hypothesis-testing procedure. According to the null hypothesis in this case,

$\mu = $ _____

$\sigma = $ _____

Answer:
$\mu = 0$
$\sigma = 3.5$

13 Using a significance level of $\alpha = 0.05$, we established a critical region of $z \leq -1.65$. If the null hypothesis is true and we repeat the same experiment a great number of times, how often would we expect to obtain a sample mean with a z score in the critical region?

Answer: 5 percent of the time

14 If the alternatives $\mu < 0$ were true, we should expect to obtain a sample mean with a z score in the critical region (more/less) often.

Answer: More

> *A different approach to the same problem would be to use $\bar{x}$ and s to establish a confidence interval for μ. The formula for the confidence interval is*

$$\bar{x} \pm z_0 \frac{s}{\sqrt{n}} = -5 \pm 1.65 \frac{3.5}{\sqrt{49}} = -5 \pm 0.825$$

> μ *is between –5.825 and –4.175 with 90% confidence.*
> *Since cases in which μ is even smaller than –5.825 also confirm the theory, you can say that $\mu \leq -4.175$ with 95% confidence. Clearly, $\mu < 0$ with at least 95% confidence. An $\bar{x}$ that is exactly on the boundary of the critical region ($z = -1.65$) will give exactly 95% confidence that $\mu < 0$.*

15 For samples larger than thirty, the method we have used is valid. For smaller samples we can use the same method with the t distribution, provided we can assume that the population of difference scores is approximately normally distributed. For example, a psychologist believes that a particular type of memory training will influence the ability of children to remember nonsense syllables. He tests a class of twenty-five children with a list of nonsense syllables before training and with an equivalent list after training, and finds the difference score for each child. Before the psychologist can use a t test on his results, he must check the assumption that the difference scores are _____ .

Answer: Approximately normally distributed.

> *The mathematical derivation of the t tables assumes that the population is normally distributed. In practice a t test will usually lead to the correct conclusion even when the distribution is fairly different from a normal distribution, provided it is not strongly skewed.*

16 If the psychologist believes that the distribution of the difference scores is likely to be very different from a normal distribution, what can he do in planning his experiment to ensure that he will be able to make a statistical test of his results?

Answer: Use a larger sample

17 Let us assume that the difference scores tend to be normally distributed in this example. Outline an appropriate statistical test below. Use a significance level of 1% and consider both ends of the sampling distribution. Use the t table (table V at the back of the book) to establish the critical region and remember that df = $n - 1$.

> Null hypothesis   ____
>
> Alternative   ____
>
> Significance level   ____
>
> Critical region   ____

Answer:

Null hypothesis	$\mu = 0$
Alternative	$\mu \neq 0$
Significance level	$\alpha = 0.01$
Critical region	$t \geq +2.80$ or $t \leq -2.80$

18 Reference formula:

$$t = \frac{\bar{x} - \mu}{s/\sqrt{n}}$$

The experimenter collects the following data: The mean difference between pretest and posttest is -5 points. The standard deviation of the sample of difference scores is 20 points.

Are the results significant?

Answer: No

$$t = \frac{-5}{20/\sqrt{25}} = \frac{-5}{4} = -1.25$$

19 Difference scores need not always be before and after scores; for example, consider the following situation: You are studying the effect of environment on intelligence test scores. You have been able to locate five sets of identical twins, one of whom was raised in an institution and the other adopted and raised in a family setting. You test all the twins and obtain the following scores:

Pair No.	Raised in Family	Institutionalized	Difference
1	105	95	−10
2	95	83	−12
3	103	103	0
4	98	96	−2
5	103	97	−6

Can you reasonably assume that these difference scores are normally distributed? There is no clear evidence that they are *not* normally distributed, and most researchers would accept the use of a *t* test. You should remember, however, that you are making this assumption on very limited evidence. Outline an appropriate statistical test for the theory that the difference in environment makes a difference in the intelligence test scores of these identical twins. Use a significance level of 5%. The *t* table is table V at the back of the book.

Null hypothesis   _____

Alternative   _____

Significance level   _____

Critical region   _____

Answer:

Null hypothesis	$\mu = 0$
Alternative	$\mu \neq 0$
Significance level	$\alpha = 0.05$
Critical region	$t \geq +2.78$ or $t \leq -2.78$

20 Reference formulas:

$$\bar{x} = \frac{\Sigma x}{n}, \quad s = \sqrt{\frac{\Sigma (x - \bar{x})^2}{n - 1}}$$

To compute *t* you will need to know the mean and standard deviation of the sample. Using the reference formulas above, compute $\bar{x}$ and *s*. (Use the square root table, table II, at the back of the book.)

$$\bar{x} = \underline{\quad\quad}$$

$$s = \underline{\quad\quad}$$

Answer:

$$\bar{x} = \frac{-30}{5} = -6.0$$

$$s = \sqrt{\frac{104}{4}} = \sqrt{26} = 5.1$$

21 Complete the *t* test by computing *t*. Is the difference between the twins' scores statistically significant?

Answer:

$$t = \frac{\bar{x} - \mu}{s/\sqrt{n}} = \frac{-6.0 - 0}{5.1/\sqrt{5}} = \frac{-6}{2.3} = 2.61$$

The difference is not statistically significant. Your statistical test does not support the theory that environment affects intelligence test scores.

22 In this experiment the result is not significant, but it is extreme enough that it would not occur by chance as often as 10 percent of the time. The sample size is small. What would you say about the chance of a Type I or Type II error in this case?

Answer: The chance of a Type II error is substantial. This statistical test may well have led us to reject a true theory.

THE DIFFERENCE BETWEEN TWO INDEPENDENT SAMPLE MEANS

The statistical analysis appropriate for comparing two independent samples is not the same as the procedure you have just learned for testing difference scores.

23 A common type of experimental study involves the comparison of an experimental sample and a control sample; for example, an experimenter interested in the effect of a plant hormone on the growth of beanstalks might treat alternate rows of beans with the hormone, leaving half the plants untreated. She will measure the height of each beanstalk and then use a statistical test to determine if the difference in mean height between the two samples of beanstalks is significant. This study does not involve the kind of individual-by-individual matching that was used in the difference score studies. Selecting one bean for the treated group does not give any information about what beans will be chosen for the other group.

Would selecting one dieter's "before" weight tell you anything about the composition of the "after" sample?

Answer: Yes. The "after" sample will include the same person.

24 In which problem above would you say the two samples are independent?

(a) The bean problem

(b) The dieting problem

Answer:

(a) The bean problem

25 For the bean problem there is a statistical test we can use to test the null hypothesis that two independent samples come from populations with the same mean. Let μ_1 be the mean height of the treated plants and μ_2 the mean height of the untreated plants. We would state the null hypothesis this way:

$$\text{Null hypothesis } \mu_1 = \mu_2$$

How would you state the alternative?

Answer:

$$\mu_1 \neq \mu_2$$

26 If the experimenter has a theory that the plant hormone will stunt the growth of the beanstalks, how would you state the alternative? Remember that μ_1 refers to the treated plants.

Answer:

$$\mu_1 < \mu_2$$

27 If both samples are large, we can compute a z score for the difference between means. If the samples are small, we can use the same techniques to compute a t score, provided certain assumptions are true. We will not explain the mathematical derivation of the formulas you will use. The end result is a z score or t score that you can compare with the appropriate table. Let us start with the case of large samples. The experimenter has a theory that predicts that the hormone will stunt the growth of beans, and she will consider only one end of the sampling distribution. She will use a significance level of 0.01. Her samples contain 100 beanstalks each. Outline an appropriate statistical test.

Null hypothesis   ——

Alternative   ——

Significance level   ——

Critical region   ——

Answer:

Null hypothesis	$\mu_1 = \mu_2$
Alternative	$\mu_1 < \mu_2$
Significance level	$\alpha = 0.01$
Critical region	$z \le -2.33$

28 The formula used for computing a z score for the difference between two means is

$$z = \frac{\bar{x}_1 - \bar{x}_2}{\sqrt{s_1^2/n_1 + s_2^2/n_2}}$$

For the following data, compute a z score:

Sample 1 (treated)	Sample 2 (untreated)
$n_1 = 100$	$n_2 = 100$
$\bar{x}_1 = 27\,\text{in.}$	$\bar{x}_2 = 29\,\text{in.}$
$s_1 = 5\text{ in.}$	$s_2 = 4\text{ in.}$

Answer:

$$z = 3.13$$

$$z = \frac{27 - 29}{\sqrt{5^2/100 + 4^2/100}} = \frac{-2}{\sqrt{25/100 + 16/100}} = \frac{-2}{\sqrt{0.4}} = -3.13$$

The experimenter's theory is supported by the statistical test. Treated beanstalks are significantly shorter.

29 Reference formula:

$$z = \frac{\bar{x}_1 - \bar{x}_2}{\sqrt{s_1^2/n_1 + s_2^2/n_2}}$$

A manufacturer suspects a difference in the quality of the spare parts he receives from two suppliers. He obtains the following data on the service life of random samples of parts from the two suppliers:

Supplier A	Supplier B
$n_1 = 50$	$n_2 = 100$
$\bar{x}_1 = 150$	$\bar{x}_2 = 153$
$s_1 = 10$	$s_2 = 5$

Outline an appropriate statistical test, using the 1% significance level, and compute z. Is the difference between the two samples statistically significant?

Answer:

Null hypothesis	$\mu_1 = \mu_2$
Alternative	$\mu_1 \neq \mu_2$
Significance level	$\alpha = 0.01$
Critical region	$z \leq +2.58$ or $z \geq -2.58$

$$z = 2.0 \text{ (not significant)}$$

$$z = \frac{150 - 153}{\sqrt{10^2/50 + 5^2/100}} = \frac{-3}{\sqrt{2 + 0.25}} = \frac{-3}{1.5} = -2.0$$

The two suppliers' parts are not significantly different.

30 To use this type of test with small samples, two assumptions must be made:

(a) The two populations involved are normally distributed.

(b) The populations have equal standard deviation; that is, $\sigma_1 = \sigma_2$.

Since the populations are unknown, you can never be certain if these assumptions are valid, but you can make some judgments about their reasonableness by looking at the data. For example, consider the following cases:

Case 1. You are testing for a significant difference between the following samples:

Sample 1   1, 4, 6, 8, 10, 11, 12, 14, 16, 18

$\bar{x} = 10, s = 5.2$

Sample 2   2, 5, 6, 8, 11, 13, 14, 15, 17, 20

$\bar{x} = 11.1, s = 5.7$

Case 2. You are testing for a significant difference between the following samples:

Sample 1   1, 4, 6, 8, 10, 11, 12, 14, 16, 18

$\bar{x} = 10, s = 5.2$

Sample 2   1, 1, 2, 2, 2, 3, 5, 10, 24, 38

$\bar{x} = 8.8, s = 12.4$

Which case better matches assumption (a) (case 1/case 2)?

Answer: Case 1. Sample 2 in case 2 is skewed.

31 Which case better matches assumption (b) (case 1/case 2)?

Answer: Case 1. The values of *s* for the two samples in case 2 are quite different; one value is more than twice as large as the other.

32 Using *t* to test the significance of the difference between the means of two independent small samples involves two assumptions:

(a) Both distributions are _____ .

(b) The population standard deviations are _____ .

Answer:

(a) Normal

(b) Equal

> *In practice, this t test will usually lead to the correct conclusion provided the two samples are equal in size and neither of the distributions is strongly skewed.*

33 The formula used for computing a t score for the difference between two means is

$$t = \frac{\bar{x}_1 - \bar{x}_2}{\sqrt{s^2/n_1 + s^2/n_2}}$$

In this formula s^2 is a pooled estimate of the population variance (σ^2) based on the s of both samples combined. The formula for computing s^2 is

$$s^2 = \frac{(n_1 - 1)\, s_1^2 + (n_2 - 1)\, s_2^2}{n_1 + n_2 - 2}$$

For the following data compute the t score:

Sample 1	Sample 2
$n_1 = 10.0$	$n_2 = 10.0$
$x_1 = 10.0$	$x_2 = 11.1$
$s_1 = 5.2$	$s_2 = 5.7$

Answer:

$$s^2 = \frac{9(5.2)^2 + 9(5.7)^2}{10 + 10 - 2} = \frac{9(5.2^2 + 5.7^2)}{18} = \frac{27.04 + 32.49}{2} = \frac{59.53}{2} = 29.765$$

$$t = \frac{10.0 - 11.1}{\sqrt{29.765/10 + 29.765/10}} = \frac{-1.1}{\sqrt{59.53/10}} = \frac{-1.1}{\sqrt{5.95}} = \frac{-1.1}{2.44} = 0.45$$

34 As you recall, when using the t table you must take into account the degrees of freedom. When you use only one sample, df = $n - 1$, but when you have two samples you must add the degrees of freedom for each sample. Then df = $(n_1 - 1) + (n_2 - 1)$, or, expressed in a simpler form, df = $n_1 + n_2 - 2$. What are the degrees of freedom for the problem in frame 33?

Answer: 18 (=10 + 10 – 2)

35 For the following statistical test, establish a critical region for the problem in frame 33 (df = 18).

Null hypothesis	$\mu_1 = \mu_2$
Alternative	$\mu_1 \neq \mu_2$
Significance level	$\alpha = 0.05$
Critical region	_____

Answer: $t \geq 2.10$ or $t \leq -2.10$

SPREADSHEET t TEST FUNCTIONS

In addition to providing you with a way to look up critical values of the normal and t distributions, spreadsheet programs provide a full-service automated t test.

When you use the t test functions, you identify the ranges that contain the two sets of sample data, the type of test you want, and if you want a one-tailed or a two-tailed test. The spreadsheet program then computes the probability of the sample result against a null hypothesis of no difference.

The formulas are:

Microsoft Excel	Lotus 1-2-3
=TTEST(range1, range2, tails, type)	@ttest(range1, range2, type, tails)

The types are:

	Microsoft Excel	Lotus 1-2-3
Paired data	Type 1	Type 2
Independent samples with approximately equal variance	Type 2	Type 0

(A third type, called heteroscedastic, applicable to situations where the assumptions of the basic t test cannot be met, is not covered in this book.)

36 For example, suppose your pretest data are in cells A10 through A25 and your posttest data are in cells B10 through B25. What formula will you use to apply a single-tailed t test to these paired data? (Answer the question for the spreadsheet program you use.) _____

Answer:

If you use Microsoft Excel, =TTEST(A10:A25, B10:B25, 1, 1)
If you use Lotus 1-2-3, @ttest(A10..A25, B10..B25, 2, 1)

37 When you enter the formula in your spreadsheet, the cell displays 0.04314. What does this number represent? _____
Are the results significant at the 5% level? At the 1% level?

Answer:

The probability of finding by chance a difference as large as the difference between the two sets of scores
The results are significant at the 5% level, but not at the 1% level.

38 There are two dart boards in a pub. You have the impression that one of the boards is easier to play than the other, but you are not sure which board is the easier one. Before you begin to investigate why this might be, you want to see if there is a significant difference between scores on the two boards. You have 27 scores from board A (in cells A11–A37) and 23 scores from board B (in cells B11–B33). What formula will you use to perform a t test? (Answer the question for the spreadsheet program you use.) _____

Answer: Since you have no hypothesis about which board is the easier one, use a two-tailed test.

If you use Microsoft Excel, =TTEST(A11:A37, B11:B33, 2, 2)
If you use Lotus 1-2-3, @ttest(A10..A25, B10..B25, 0, 2)

CHOOSING THE RIGHT TEST

Using an inappropriate statistical test will give misleading results. To select an appropriate statistical test you must consider two questions:

(a) What are your null hypothesis and alternative?

(b) What assumptions and limitations apply to the statistical tests you might use?

39 Let us look at some examples. A reading test is supplied by a national publisher. The test is accompanied by a manual that describes the results of the administration of the test to some 4,000 tenth-grade students in various parts of the country. According to this manual, the distribution of scores on the test was approximately normal, with a mean of 30 and a standard deviation of 5. You wish to know if the performance of students in your city differs significantly from that of the 4,000 students described in the manual, so you select a random sample of 25 tenth-grade students and test them. What are your null hypothesis and alternative for the statistical test?

Null hypothesis   _____

Alternative   _____

Answer:

Null hypothesis $\mu = 30$
Alternative $\mu \neq 30$

40 The form of your null hypothesis tells you what statistical tests are available. Which of the following corresponds to your null hypothesis in this case?

(a) $\mu = C$, where C is a known constant that you could state before the experiment; for example, $\mu = 0$ or $\mu = 10$.

(b) $\mu_1 = \mu_2$, where μ_1 and μ_2 are both estimated on the basis of samples.

Answer:

(a) μ = C

41 Look at the summary of formulas in table I at the back of the book. What two formulas are listed for testing the hypothesis μ = C?

Answer:

$$z = \frac{\bar{x} - C}{s/\sqrt{n}}$$

$$t = \frac{\bar{x} - C}{s/\sqrt{n}}$$

42 The form of your alternative tells you whether your test is "one-tailed" or "two-tailed." A two-tailed test is a test in which you take into account both ends of the sampling distribution when you set your critical region. It corresponds to an alternative such as μ ≠ C. A one-tailed test is concerned with only one end of the sampling distribution. It corresponds to an alternative such as μ ≥ C or μ ≤ C. Consider your alternative in this problem. What type of test will you use?

Answer: Two-tailed. You will reject the null hypothesis if the sample mean is either unusually low or unusually high.

43 Now ask yourself the second question: "What assumptions and limitations apply to the statistical tests you might use?" Use of a z score involves an assumption about sample size. What is it?

Answer: The sample size is greater than thirty.

44 Using a t score involves an assumption about the distribution of your unknown population. What is it?

Answer: The population is approximately normally distributed.

45 The sample size in this problem is twenty-five. Suppose you test the twenty-five students and find that their scores appear to be roughly normally distributed, with $\bar{x} = 26$ and $s = 6$. What will you use for a statistical test?

(a) A z score

(b) A t score

(c) Neither

Answer:

(b) A t score

46 Using a 1% significance level, complete the statistical test:

Null hypothesis $\mu = 30$

Alternative $\mu \neq 30$

Significance level $\alpha = 0.01$

Critical region $t >$ _____ or $t <$ _____

$t =$ _____ . The results are (significant/not significant).

Answer: Critical region: $t \geq +2.797$ or $t \leq -2.797$ (df = 24)

$$t = \frac{26 - 30}{6/\sqrt{25}} = \frac{-4}{6/5} = \frac{-20}{6} = -3.33$$

Significant
Your sample does differ significantly from the sample described in the manual.

47 Using the same reading test with a sample of twenty-five students from a different city, you find that their scores are bimodally distributed, as indicated in the illustration. There are a substantial number of students in the city who have learned English as a second language.

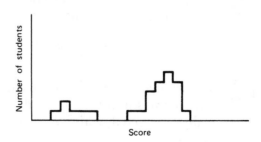

FIGURE 5-1:
Reading test scores

Again, you wish to know if the mean performance of all students in this city differs significantly from that of the 4,000 students described in the test manual. What will you use for a statistical test in this case?

(a) A z score

(b) A t score

(c) Neither

Answer:

(c) Neither. In this case you cannot assume that the sample comes from a normally distributed population, and the sample is too small to use a z score. One solution would be to obtain a larger sample so that a z score could be used. Perhaps a more appropriate solution would be to treat the students who learned English as a second language as a separate population and draw two separate samples.

48 All tenth-grade students in your city have been tested. This population consists of 3,000 tenth-graders. Their reading scores are distributed as shown below. A group of 100 from this population has been selected for a special enrichment program, and you suspect that the selection procedures favored the better readers. To check your theory, you obtain the reading scores of the 100 students selected for the special program. What are the null hypothesis and alternative?

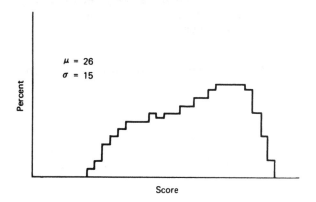

FIGURE 5-2:
Reading test scores

Answer:

Null hypothesis $\mu = 26$
Alternative $\mu > 26$

49 Which of the following corresponds to your null hypothesis in this case?

(a) $\mu = C$

(b) $\mu_1 = \mu_2$

Answer:

(a) $\mu = C$

50 The form of the null hypothesis indicates that statistical tests are available with z scores or t scores. The form of the alternative indicates that the test will be (one-tailed/two-tailed).

Answer: One-tailed

51 What assumptions and limitations apply to the available tests? For a z score? For a t score?

Answer:

For a z score, n must be greater than 30.
For a t score, the sample is assumed to have come from a normally distributed population.

52 Considering these assumptions, what will you use for a statistical test in this case?

(a) A z score

(b) A t score

(c) Neither

Answer:

(a) A z score. The sample is large enough to use z.

Although the population is not normally distributed, most researchers would probably accept conclusions based on a t score for a smaller sample in this case.

53 Theoretical analysis leads a researcher to predict that the trout in one lake (Deep Lake) will have higher concentrations of chlorinated hydrocarbons in their livers than the trout in another lake (Blue Lake). He catches ten fish from each lake. What are the null hypothesis and alternative?

Answer:

Null hypothesis $\mu_1 = \mu_2$
Alternative $\mu_1 > \mu_2$

54 Look at the summary of formulas in table I at the back of the book. What formulas are listed for testing this hypothesis?

Answer:

$$z = \frac{\bar{x}_1 - \bar{x}_2}{\sqrt{s_1^2/n_1 + s_2^2/n_2}}$$

$$t = \frac{\bar{x}_1 - \bar{x}_2}{\sqrt{s^2/n_1 + s^2/n_2}}$$

55 The statistical test will be (one-tailed/two-tailed)?

Answer: One-tailed

56 What assumptions apply to the z score test in this case?

Answer: Both samples are larger than thirty.

57 What assumptions apply to the t score test in this case?

Answer: The populations are normally distributed and have equal standard deviations.

58 The two samples appear to be more or less normally distributed. The following amounts of chlorinated hydrocarbons are detected:

Sample 1 (Deep Lake)	Sample 2 (Blue Lake)
$\bar{x} = 8$	$\bar{x} = 4$
$s = 3$	$s = 2$
$n = 10$	$n = 10$

Review the problem as outlined in frame 53 and the following frames and complete an appropriate statistical test. Do these results confirm the researcher's theory? (Use the 1% significance level.)

Answer:

Null hypothesis	$\mu_1 = \mu_2$
Alternative	$\mu_1 > \mu_2$
Significance level	$\alpha = 0.01$
Critical region	$t \geq +2.55$ (df = 18)

$$s^2 = \frac{9(2)^2 + 9(3)^2}{18} = 6.5 \quad t = \frac{8-4}{\sqrt{1.3}} = 3.51$$

The researcher's theory is confirmed. The t test is used because the samples are small and s_1 and s_2 are not drastically different.

59 It is important to distinguish between independent samples and samples that are actually paired data and should be used to compute difference scores. For example, a statistically minded shopper wants to compare the prices of the two supermarkets in his neighborhood. From a list of all types of canned and frozen goods that are available, he randomly selects 43 different items. He then determines the price of each of these items at each store, so that he has a list of 43 prices for each of the two stores. Do his two lists of prices constitute independent samples or should he compute difference scores?

Answer: He should compute difference scores because he priced the same items at each store.

60 What are an appropriate null hypothesis and alternative for a statistical test?

Answer:

Null hypothesis	$\mu = 0$
Alternative	$\mu \neq 0$

61 What formula should he use for a statistical test?

Answer:

$$z = \frac{\bar{x} - C}{s/\sqrt{n}}$$

62 An economist is investigating price differences between suburban and central city stores. She determines the total price of a standard shopping list at a random sample of 18 suburban stores and 22 central city stores, so that she has 18 shopping-list totals from suburban stores and 22 shopping-list totals from central city stores. Has she two independent samples?

Answer: Yes. Selecting a given suburban store has no influence on what central city stores will be selected.

63 Outline an appropriate statistical test.

       Null hypothesis   _____

       Alternative   _____

       Significance level   0.05

       Critical region   _____

       Formula for test   _____

Answer:

Null hypothesis $\mu = 0$
Alternative $\mu \neq 0$
Critical region $t \geq +2.02$ or $t \leq -2.02$ (df = 38)

$$\text{Formula for test: } t = \frac{\bar{x}_1 - \bar{x}_2}{\sqrt{s^2/n_1 + s^2/n_2}}$$

$$s^2 = \frac{(n_1 - 1)s_1^2 + (n_2 - 1)s_2^2}{n_1 + n_2 - 2}$$

A last caution about one-tailed versus two-tailed tests. A one-tailed test is appropriate only when you have clearly decided in advance of the experiment that results in the opposite direction from your alternative are of no interest whatsoever. You cannot look at the data and then decide to use a one-tailed test. For this reason, many researchers always use two-tailed tests, even when they have formed a tentative conclusion about the direction of the results. When in doubt, use a two-tailed test.

SELF-TEST

If you have successfully completed this chapter, you can now set up formal statistical tests of hypotheses about differences between the means of two sets of data. You can:

- Test the significance of a set of difference scores

- Test the significance of the difference between the means of two independent samples

- Recognize when you have independent samples, and when you have paired data, which result in difference scores

Now try these review problems. Table I at the back of the book lists any formulas you may need for reference.

1. You are studying problem-solving performance, using time as a measure. With the kind of problem you are using, the distributions you obtain are typically bimodal. Your subjects either solve the set of problems quickly or take a long time; only rarely do they take an intermediate amount of time. You select a sample of ten subjects and give them instruction that you believe will reduce their mean time for solving the problem set. You wish to compare their performance with that of another sample of ten who did not receive the instruction. Outline an appropriate statistical test and suggest changes in the plan of the experiment if necessary. Use a 5% significance level.

2. A laboratory owns two precision measuring devices. The director suspects that there is a slight difference in calibration between the two, so that one of them (he doesn't know which) tends to give slightly higher readings than the other. He proposes to check the two devices by taking readings of 50 objects on both machines. Thus he will have readings of objects 1 through 50 on machine A and readings of objects 1 through 50 on machine B. Outline an appropriate test at the 5% significance level and suggest changes in the plan if necessary.

3. From appropriately selected samples, two sets of IQ scores are obtained, as summarized below:

Group 1	Group 2
$n = 16$	$n = 14$
$\bar{x} = 107$	$\bar{x} = 112$
$s = 10$	$s = 8$

Is there a significant difference between the two groups? Use the 5% significance level.

ANSWERS

To review a problem, study the frames indicated after the answer.

1. You cannot perform a statistical test as the experiment is planned. The samples are too small to use a z test, and to use a t test you must be able to assume that the populations are approximately normally distributed. Your experience indicates that you cannot make this assumption. The solution is to use samples of thirty or more. With larger samples, the statistical test is the following:

Null hypothesis	$\mu_1 = \mu_2$
Alternative	$\mu_1 < \mu_2$
Significance level	$\alpha = 0.05$
Critical region	$z \leq -1.65$

The formula for z is

$$z = \frac{\bar{x}_1 - \bar{x}_2}{\sqrt{s_1^2/n_1 + s_2^2/n_2}}$$

See frames 23 to 35 and 39 to 63.

2. You must use difference scores in this case because the samples are not independent. Since you have 50 difference scores, use z.

Null hypothesis	$\mu = 0$
Alternative	$\mu \neq 0$
Significance level	$\alpha = 0.05$
Critical region	$z \leq -1.96$ or $z \geq +1.96$

The formula for z is

$$z = \frac{\bar{x} - C}{s/\sqrt{n}}$$

See frames 1 to 22 and 39 to 63.

3. The samples are small, so you must use t. In the absence of evidence to the contrary, it is reasonable to assume that the populations of IQ scores are approximately normally distributed.

Null hypothesis $\mu_1 = \mu_2$

Alternative $\mu_1 \neq \mu_2$

Significance level $\alpha = 0.05$

Critical region $t \geq +2.05$ or $t \leq -2.05$ (df = 16 + 14 - 2 = 28)

$$\text{Formula for test: } t = \frac{\bar{x}_1 - \bar{x}_2}{\sqrt{s^2/n_1 + s^2/n_2}}$$

$$s^2 = \frac{(n_1 - 1)s_1^2 + (n_2 - 1)s_2^2}{n_1 + n_2 - 2}$$

$$s^2 = \frac{15(100) + 13(64)}{16 + 14 - 2} = \frac{1500 + 832}{28} = 83.3$$

$$t = \frac{107 - 112}{\sqrt{83.3/16 + 83.3/14}} = \frac{-5}{\sqrt{5.21 + 5.95}} = \frac{-5}{3.35} = -1.49$$

The difference between the two groups is not significant. *See frames 30 to 35 and 39 to 63.*

6 The Difference between Two Variances or Several Means

The difference between two variances can be studied using another sampling distribution called the *F* distribution. The approach should, by now, be familiar. You compute a value for *F* using the data from your samples and a sampling distribution table to determine if the value is in the critical region.

A related technique called *analysis of variance* allows you to consider data from several samples at the same time. You try to distinguish systematic differences between sample groups from the chance variation found within each group. Analysis of variance requires you to compute an *F* value.

When you have completed this chapter, you will be able to:

- Test the significance of the difference between two sample variances

- Perform an analysis of variance to determine if the differences among the means of a group of samples are statistically significant

- Determine if given data meet the assumptions of these tests

COMPARISON OF TWO VARIANCES

Just as two means can be compared by using the sampling distribution of *t*, two variances can be compared by using the sampling distribution of *F*. The procedure is very simple.

1 The variability of a population or a sample can be described by its standard deviation or by its variance. The variance is simply the square of the standard deviation, σ; for example, if σ for a population is 6, the variance of the population is _____ .

Answer: 36

2 If s for a sample is 5, the variance of the sample is _____ .

Answer: 25

3 The formula for the F ratio is

$$F = \frac{s_1^2}{s_2^2}$$

The usual way to compare the variability of two samples is to divide the larger variance by the smaller variance. The result of this computation is called an F ratio. For example, if the variance of one sample is 25 and the variance of another sample is 4, the F ratio is 25/4 or 6.25. If the variance of one sample is 36 and the variance of another is 9, the F ratio is _____ .

Answer: 36/9 = 4.0

4 If one sample has a variance of 5 and the other a variance of 25, the F ratio is

Answer: 25/5 = 5.0

5 An F ratio close to 1.0 indicates that the two samples have (similar/different) variances.

Answer: Similar

6 Two samples with very different variances would result in an F ratio that is (close to 1.0/large).

Answer: Large

7 In computing an F ratio, you normally divide the (larger/smaller) variance by the (larger/smaller) variance.

Answer:

Larger
Smaller

8 If two samples are drawn at random from populations with the same variance, their F ratio is most likely to be close to 1.0. A large F ratio will be relatively unlikely. For normally distributed populations it is possible to construct theoretical sampling distributions for F. This fact can be used to perform a statistical test of a hypothesis about the difference between the variance of two samples. For example, a researcher believes that stress will increase the variability of a sample's test scores. He administers the same test to two groups of twenty-five college applicants. One group is told that this test will be important in determining whether they will be admitted. The other group is told that the test is being administered for research purposes only. His statistical test is as follows:

Null hypothesis	$\sigma_1 = \sigma_2$
Alternative	$\sigma_1 > \sigma_2$
Significance level	$\alpha = 1\%$
Critical region	$F > 2.66$ (This value is obtained from a table, as you will soon see.)

His results are as follows:

Group 1 (stress)	Group 2 (no stress)
$n = 25$	$n = 25$
$\bar{x} = 120$	$\bar{x} = 110$
$s = 20$	$s = 10$
$s^2 = 400$	$s^2 = 100$

Compute F. _____ .

Answer: $F = 4.0$

9 Do the results support the researcher's theory?

Answer: Yes

10 The tables used to establish a critical region for this test will tell you that a given value of F would occur by chance no more than 5 percent of the time, for example, or no more than 1 percent of the time. Table VI at the back of the book is such a table. Look at it. What is its title?

Answer: "Critical Points of the F Distribution"

11 To use the F table you must know the degrees of freedom for both s_1^2 and s_2^2. As in the t test, df $= n - 1$.
As an example,

$$s_1^2 = 36 \qquad s_2^2 = 25$$

$$n_1 = 15 \qquad n_2 = 21$$

The degrees of freedom for s_1^2 are 14. The degrees of freedom for s_2^2 are
_____ .

Answer: 20

12 Look at the F table and locate the area on the table that corresponds to 14 degrees of freedom for s_1^2 and 20 degrees of freedom for s_2^2. You should find two values of F in this area. What are they?

Answer: 2.23 and 3.13

13 According to the table, how often would you expect an F of 3.13 or greater to occur by chance with these sample sizes?

Answer: One percent of the time

14 Five percent of the time, chance would produce an F greater than _____ .

Answer: 2.23

15 Assume for a moment that you are operating a frozen grapefruit juice factory. To buy fruit at favorable prices you must buy the grower's entire crop, but unusually large or unusually small fruit can jam your machinery and must be eliminated before squeezing. For this reason you would like to buy crops that are relatively uniform in size; that is, you prefer crops with a (large/small) variance.

Answer: Small

16 Two growers offer you their crops. Grower A asks a slightly higher price than grower B, but he says that his grapefruit are more uniform in size. To check this assertion you ask for a random sample of each crop. Each grower sends you a crate of twenty-five grapefruit. You measure the grapefruit in each sample and obtain the following information:

(a) The size of the fruit is approximately normally distributed for both samples.

(b) For grower A the mean diameter of the fruit is 4.5 in., with a standard deviation of 0.5 in.

(c) For grower B the mean diameter of the fruit is 4.5 in., with a standard deviation of 1 in.

State an appropriate null hypothesis and alternative for a statistical test.

Answer:

Null hypothesis $\sigma_1 = \sigma_2$
Alternative $\sigma_1 < \sigma_2$

17 For a significance level of 5%, what is the critical region? (Use the table.)
$F \geq$ _____ .

Answer: 1.98

18 Compute F. Are the results significant?

Answer:

$$F = \frac{1^2}{0.5^2} = \frac{1}{0.25} = 4.0$$

The results are significant. The fruit from grower A is more uniform in size.

19 In the cases you have considered so far, the alternative called for a one-tailed statistical test; that is, your theory specified which of the two variances would be larger. Sometimes your alternative is two-tailed. It says only that $\sigma_1 \neq \sigma_2$ without stating which one will be larger. In this case you will use whichever variance happens to be larger as the numerator of the F ratio, but you must double the probabilities when you use the F table. The 5% values become 10% values and the 1% value become _____ values.

Answer: 2%

20 You have been given the following statistics on two samples. From experience you are quite certain that the populations are normally distributed, but you suspect that the two populations are not equally variable.

Sample 1	Sample 2
$n = 10$	$n = 10$
$x = 8.8$	$x = 12.7$
$s = 12$	$s = 5$

What would be an appropriate null hypothesis and alternative for investigating this matter?

Answer:

Null hypothesis $\sigma_1 = \sigma_2$
Alternative $\sigma_1 \neq \sigma_2$

21 Using a 2% significance level, establish a critical region.

Answer: Critical region: $F \geq 5.35$

22 Compute F.

Answer:

$$F = \frac{12^2}{5^2} = \frac{144}{25} = 5.76$$

23 Will a *t* test be appropriate in this case?

Answer: No. The difference between the variance of the two samples is exceptionally large. It is not likely that they came from populations with the same variance.

USING A SPREADSHEET TO LOOK UP CRITICAL VALUES OF *F*

Spreadsheet functions for the *F* distribution are similar to the other spreadsheet statistical functions. There are functions that return the probability of *F* equal to or greater than a given value. And there are inverse functions that return the critical value of *F* that corresponds to a given significance level.

The *F* distribution is not symmetrical like *z* and *t*. Because *F* is a ratio between two variances, its value can never be negative. The critical region consists of large positive values of *F*.

Look at Figure 6-1. The shaded area under the curve is the probability referred to by the spreadsheet functions. The spreadsheet programs return a probability appropriate for a one-tailed test, in the same way that most printed tables do.

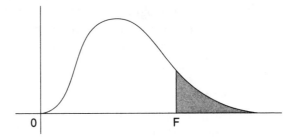

FIGURE 6-1:
The *F* distribution

Since the F distribution depends on degrees of freedom, you must always include degrees of freedom for the numerator and the denominator. For Microsoft Excel, there are two different functions, FDIST and FINV. For Lotus 1-2-3, there is one function name, but there are two types: type 0 returns a critical value of F and type 1 returns a probability.

	Microsoft Excel	Lotus 1-2-3
To find a probability	=FDIST(F, dfnum, dfdenom)	@fdist(F, dfnum, dfdenom, 0)
To find a critical value of F	=FINV(p, dfnum, dfdenom)	@fdist(p, dfnum, dfdenom, 1)

24 Write the formula you would enter into a cell of your spreadsheet to have your program calculate the F score that corresponds to a 3% significance level with 7 and 6 degrees of freedom. (Answer the question for the spreadsheet program you use.) _____

Answer:

If you use Microsoft Excel, =FINV(0.03, 7, 6)
If you use Lotus 1-2-3, @fdist(0.03, 7, 6,1)

25 Write the formula you would enter into a cell of your spreadsheet to have your program calculate the probability of an F score of 3.9174 with 2 degrees of freedom in the numerator and 12 degrees of freedom in the denominator. (Answer the question for the spreadsheet program you use.) _____

Answer:

If you use Microsoft Excel, =FDIST(3.9174, 7, 6)
If you use Lotus 1-2-3, @fdist(3.9174, 7, 6, 0)

ANALYSIS OF VARIANCE

The F ratio can be used to test the null hypothesis that a number of samples all come from populations with the same mean. The procedure used for this test is called analysis of variance.

26 In the analysis of variance we use the difference among sample means to estimate the variance of the population. We also make a separate estimate of the population variance based only on the differences among individuals within each sample. If the samples all come from populations with the same mean, the differences between sample means will be relatively (large/small).

Answer: Small

27 If the differences between sample means are relatively large, we will conclude that the mean of the populations (is/is not) the same.

Answer: Is not

28 The estimate of variance based on the differences between the means of groups is called "between-groups variance." The estimate of variance based only on the differences between individuals is called "within-groups variance," or "error variance." To reject the null hypothesis, between-groups variance must be large compared with ____.

Answer: Within-groups variance or error variance

29 Let us see how between-groups variance and within-groups variance can be estimated. The purpose of this explanation is to show you the basic logic of the test. You will learn simplified procedures for computation later. A researcher believes that the color of a toy will affect how long children will play with it. From a population of preschool children she obtains four sample groups of ten children each. Using the same stuffed animal, but in a different color for each sample group, she observes how many minutes each child in each sample group spends playing with the toy during a 10-minute session. Her null hypothesis is that $\mu_1 = \mu_2 = \mu_3 = \mu_4$. The alternative is that not all the means are equal; that is, that the color of the toy does make a difference. Her data look like this:

Group 1 (red giraffe)	Group 2 yellow giraffe)	Group 3 (green giraffe)	Group 4 (blue giraffe)
1	3	2	5
2	2	4	3
5	6	2	1
7	3	1	2
6	2	2	1
1	8	3	3
2	7	4	4
2	5	1	2
4	6	3	3
4	8	2	1
$n_1 = 10$	$n_2 = 10$	$n_3 = 10$	$n_4 = 10$
$\bar{x}_1 = 3.4$	$\bar{x}_2 = 5.0$	$\bar{x}_3 = 2.4$	$\bar{x}_4 = 10$
$s_1^2 = 4.5$	$s_2^2 = 5.6$	$s_3^2 = 1.2$	$s_4^2 = 1.8$

All groups combined:

$$N = 40$$

$$\bar{x}_T = 3.3$$

$$s_T^2 = 4.1$$

The between-groups variance estimate is based on the differences between (individuals/group means).

Answer: Group means

30 Consider the distribution of the four group means. This distribution is

(a) A population distribution

(b) A sample distribution

(c) A sampling distribution

Answer:

(c) A sampling distribution

31 As you recall from our discussion of the central limit theorem,

$$\sigma_{\bar{x}} = \frac{\sigma}{\sqrt{n}}$$

That is, the standard deviation of the sampling distribution of x depends on two things, σ and n.

What does σ stand for?

What does n stand for?

Answer:

σ is the standard deviation of the population.
n is the sample size.

32 At the moment we are interested in computing s^2, an estimate of the variance of the population. We can find it by using a formula based on the formula above. Using s in the place of σ to indicate that this is an estimate, we can say $s_{\bar{x}} = s/\sqrt{n}$. Then $s_{\bar{x}}^2 = s^2/n$, and $s^2 = n s_{\bar{x}}^2$. What is the size of the samples that make up our sampling distribution? $n =$ _____

Answer: 10

33 $s_{\bar{x}}^2$ is simply the variance of the following numbers:

3.4, 5.0, 2.4, and 2.5.

What are these numbers?

Answer: The means of the four samples.

34 The variance of the sample means works out to be 1.45. (If you would like practice, check it for yourself.) On the basis of this information, what is the variance of the total population?

$$s = n s_{\bar{x}}^2 = \underline{\hspace{1cm}}.$$

Answer: 10(1.45) = 14.5

35 14.5 is the (between/within) groups variance estimate.

Answer: Between

36 The between-groups variance estimate is based on the variance of (individual scores/group means).

Answer: Group means

37 The within-groups variance estimate is based on the variances of the individual scores within sample groups. What are the variances of individual scores within the sample groups?

$$s_1^2 = \underline{\hspace{1cm}}$$

$$s_2^2 = \underline{\hspace{1cm}}$$

$$s_3^2 = \underline{\hspace{1cm}}$$

$$s_4^2 = \underline{\hspace{1cm}}$$

Answer:

$s_1^2 = 4.5$

$s_2^2 = 5.6$

$s_3^2 = 1.2$

$s_4^2 = 1.8$

38 Are these variances affected by the differences between groups?

Answer: No

39 The variance of each of these groups is an estimate of the total population variance. To obtain an even better estimate we can average all four group variances. This is our within-groups variance estimate. Compute the within-groups variance estimate for this example.

Answer:

$$\frac{4.5 + 5.6 + 1.2 + 1.8}{4} = 3.28$$

40 Compute the F ratio.

$$F = \frac{\text{between-groups variance}}{\text{within-groups variance}} = \underline{\hspace{1cm}}$$

Answer:

$$F = \frac{14.5}{3.28} = 4.42$$

41 To evaluate this F ratio we still need to know the degrees of freedom for the between-groups and within-groups variances. The between-groups variance estimate is based on four means. Degrees of freedom for this variance are $4 - 1 = 3$. The within-groups variance is an average of several variances, and the degrees of freedom for it are the total of the degrees of freedom for each of the groups. What are the degrees of freedom for one group of 10?

Answer: 9

42 What are the total degrees of freedom for four groups of 10?

Answer: $4 \times 9 = 36$

43 The degrees of freedom for s_b^2 (between-groups variance) are _____ . The degrees of freedom for s_w^2 (within-groups variance) are _____ . For a 1% significance level the critical region is $F \geq$ _____ .

Answer:

3
36
4.38

44 Will the experimenter reject the null hypothesis?

Answer: Yes, because the F ratio is in the critical region. She can conclude that the color of the toy does make a difference in its attractiveness.

45 In the example you have just considered, the sample groups were of equal size. When the sample groups are not of equal size, it is necessary to weight the means and variances of the sample groups according to their sizes. The computations are more easily done when you use "sums of squares" instead of means and variances. A sum of squares is the numerator (upper term) of a variance estimate. Circle the part of the following formula that represents the sum of squares.

$$s = \frac{\sum (x - \bar{x})^2}{n - 1}$$

Answer:

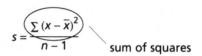

$$s = \frac{\sum (x - \bar{x})^2}{n - 1}$$ sum of squares

46 The between-groups sum of squares is the sum of squares you will use to compute the between-groups variance estimate. The within-groups sum of squares is the sum of squares you will use to compute the within-groups variance estimate. If you consider all the groups combined as a single large sample, you can also compute a total sum of squares. To compute the variance of all the observations combined as a single sample, you would use the _____ sum of squares.

Answer: Total

47 Adding the between-groups sum of squares to the within-groups sum of squares will always give you the total sum of squares. This fact can simplify computations. For example, if you know that the total sum of squares is 75 and the between-groups sum of squares is 25, the within-groups sum of squares must be _____ .

Answer: 50

In practical work the mathematical procedures for analysis of variance are normally done by computer. To understand the terminology and to appreciate what you are missing, you nevertheless ought to work through a few problems by hand. If you have no tolerance for computation, at least read through the explanations of the following problems step by step—or set up the detailed individual steps in a spreadsheet. You will need this background to understand computer output.

48 To perform an analysis of variance of data from varying size groups, start by setting up a table like this:

Group 1		Group 2		Group 3		Group 4	
x	x^2	x	x^2	x	x^2	x	x^2
3	9	2	4	3	9	5	25
3	9	4	16	5	25	6	36
5	25	4	16	5	25	6	36
6	36	6	36	6	36	7	49
8	64	9	81	7	49		
		10	100	7	49		
				8	64		
25	143	35	253	41	257	24	146
$n_1 = 5$		$n_2 = 6$		$n_3 = 7$		$n_4 = 4$	
$\sum x_T = 125$		$\sum x_T^2 = 799$		$N = 22$			

From the above table you can read all the information you will need to complete the analysis of variance. Notice that there are two columns for each group. The first column lists the scores for the group. The second column lists _____ .

Answer: The scores squared

49 Do you see any columns for $(x - \bar{x})$ or $(x - \bar{x})^2$?

Answer: No

50 What symbol is used to indicate the number of observations in group 1?

Answer: n_1

51 What symbol is used to indicate the total number of observations in all groups combined?

Answer: N

52 What symbol is used to indicate the grand total of all the observations?

Answer: Σx_T

53 You can compute all the necessary sums of squares, using only Σx and Σx^2 for the various groups. The formulas are as follows.

Total sum of squares:

$$\Sigma x_T^2 - \frac{(\Sigma x_T)^2}{N}$$

Degrees of freedom for the total sum of squares depend on the total size of all samples combined: df = $N - 1$.

Between-groups sum of squares:

$$\frac{(\Sigma x_1)^2}{n_1} + \frac{(\Sigma x_2)^2}{n_2} + \cdots - \frac{(\Sigma x_T)^2}{N}$$

Degrees of freedom for the between-groups sum of squares depend on the number of groups. We use g to indicate the number of groups: df = $g - 1$

Within-groups sum of squares:

Total sum of squares – between-groups sum of squares, or

$$\Sigma x_1^2 - \frac{(\Sigma x_1)^2}{n_1} + \Sigma x_2^2 - \frac{(\Sigma x_2)^2}{n_2} + \cdots$$

Degrees of freedom for the within-groups sum of squares depend on the degrees of freedom for each of the individual groups. df = $(n_1 - 1) + (n_2 - 1) + \ldots$. A simpler way to calculate this is df = $N - g$.
    In the example above, what value from the table corresponds to Σx_T?

Answer: 125

54   What value corresponds to Σx_T^2?

Answer: 799

55   What value will correspond to $(\Sigma x_T)^2$?

Answer: 15,625 (= 125^2)

56   What value corresponds to Σx_3^2?

Answer: 257

57   What value corresponds to $(\Sigma x_1)^2$?

Answer: 625 (= 25^2)

58   Calculate the total sum of squares for the data in frame 48. Refer back to the formulas if you wish.

Answer:

$$799 - \frac{(125)^2}{22} = 799 - 710.2 = 88.8$$

59   Calculate the between-groups sum of squares.

Answer:

$$\frac{(25)^2}{5} + \frac{(35)^2}{6} + \frac{(41)^2}{7} + \frac{(24)^2}{4} - \frac{(125)^2}{22} = 125 + 204.2 + 240.1 + 144 - 710.2 = 3.1$$

60   Calculate the within-groups sum of squares.

Answer:

$$88.8 - 3.1 = 85.7$$

61   Complete the following table:

	Sum of Squares	df	Variance Estimate
Total			—
Between groups			
Within groups			

Answer:

	Sum of Squares	df	Variance Estimate
Total	88.8	21	—
Between groups	3.1	3	1.03
Within groups	85.7	18	4.76

62 Reference formula:

$$F = \frac{s_b^2}{s_w^2}$$

What is the F ratio for these data? $F =$ _____ .

Answer:

$$\frac{1.03}{4.76} = 0.22$$

63 Let's review the meaning of between-groups and within-groups variances. The between-groups variance estimate is based on the means of the groups. The within-groups variance estimate is based on the variances of the individual groups. Which variance estimate reflects *only* the chance variations involved in drawing a sample?

Answer: Within-groups. The between-groups variance also reflects the intentional difference between groups.

64 Which variance estimate reflects the intentional differences between groups?

Answer: Between-groups

65 If the between-groups variance is less than the within-groups variance, you can say that the differences between groups are (large/small) compared with the chance variation involved in drawing a sample.

Answer: Small

66 What does the within-groups variance estimate reflect?

Answer: Chance, the chance variations involved in drawing a sample.

67 What does the between-groups variance estimate reflect?

Answer: The differences between groups

68 If s_b^2 is smaller than s_w^2, can the results be significant?

Answer: No

69 The following scores represent the performance on a troubleshooting test of small random samples of people with different educational backgrounds. Does educational background appear to make a difference in the performance on this test? Use a significance level of 1%.

	Scores
Group 1 (high school)	1, 3, 4
Group 2 (technical institute)	4, 5, 6, 6, 7, 8, 9
Group 3 (college)	2, 3, 3, 4

Suggestions:

(a) Set up a table for the data like the one in frame 48.

(b) Compute the sums of squares and the variance estimates and summarize them in a table like the one in frame 61.

(c) Calculate F and refer to the table to determine if it is significant.

Answer:

Group 1		Group 2		Group 3	
x	x^2	x	x^2	x	x^2
1	1	4	16	2	4
3	9	5	25	3	9
4	16	6	36	3	9
		6	36	4	16
		7	49		
		8	64		
		9	81		
8	26	45	307	12	38
	$n_1 = 5$		$n_2 = 6$		$n_3 = 4$
	$\Sigma x_T^2 = 65$		$\Sigma x_T^2 = 371$		$N = 14$

	Sum of Squares	df	Variance Estimate
Total	69.2	13	—
Between groups	44.8	2	22.40
Within groups	24.4	11	2.22

$$F = \frac{22.40}{2.22} = 10.1$$

The null hypothesis is rejected. Educational background does make a difference.

SPREADSHEET F TEST FUNCTIONS

With not too much effort, you could set up a spreadsheet to implement all the steps in an analysis of varience as we have outlined in this chapter. To make it even easier, packaged functions are provided that perform all the steps for some common situations.

Both Microsoft Excel and Lotus 1-2-3 include F test functions, which perform an analysis of variance to compare the variance of two independent samples. These functions require you to identify the two ranges containing the data you want to compare; they return a probability value in the cell you have selected.

When you use the F test functions, you identify the ranges that contain the two sets of sample data. The spreadsheet program then computes the one-tailed probability of the sample result.

The formulas are:

Microsoft Excel	Lotus 1-2-3
=FTEST(range1, range2)	@ftest(range1, range2)

70 Suppose your group A data are in cells A10 through A25 and your group B data are in cells B10 through B35. What formula will you use to test the hypothesis that the two groups have different variances? (Answer the question for the spreadsheet program you use.) _____

Answer:

If you use Microsoft Excel, =FTEST(A10:A25, B10:B35)
If you use Lotus 1-2-3, @ftest(A10..A25, B10..B35)

71 When you enter the formula in your spreadsheet, the cell displays 0.09871. What does this number represent? _____
Are the results significant at the 5% level? At the 1% level?

Answer:

The probability of finding by chance an F ratio as large as the one found for the two sets of scores.
The results are not significant at the 5% level, nor, of course, at the 1% level.

In addition to the FTEST function, Microsoft Excel provides you with tools that allow you to fill in a dialogue to perform some typical analysis of variance procedures. These tools are available under the Tools|Data Analysis menu selection.

Caution: When you use the Microsoft Excel Data Analysis tools, the detailed calculations do not appear on the spreadsheet and the result does not update automatically when input numbers change. The result of using the Data Analysis tools is simply a text printout of the analysis results in a selected area of the spreadsheet. If you use these tools, it is a good idea to include some formulas that do update automatically (for example, a calculation of cell means) as a cross-check that the data analysis printout does not require updating.

72 Look at this printout of an analysis of variance calculated by computer.

ANOVA: Single Factor

SUMMARY

Groups	Count	Sum	Average	Variance
Group A	5	25	5	4.5
Group B	6	35	5.8333	9.7667
Group C	7	41	5.8571	2.8095
Group D	4	24	6	0.6667

ANOVA

Source of Variation	SS	df	MS	F	P-value	F crit
Between groups	3.0822	3	1.0274	0.2158	0.8841	3.1599
Within groups	85.6905	18	4.7606			
Total	88.7727	21				

Note the degrees of freedom; how many groups were used in the experiment? _____

How many individuals were studied? _____

Are the differences among groups significant at the 1% level? _____

Answer:

4 groups
22 individuals
No, the results are not significant.

WHEN TO USE ANALYSIS OF VARIANCE

Certain assumptions apply to the analysis of variance. Only when these assumptions are met is it appropriate to use analysis of variance to test your hypothesis.

73 Analysis of variance and the *t* test for the difference between two means are based on similar mathematical derivations. They depend on the same assumptions and they yield the same results when applied to two samples. Both tests require what two assumptions?

Answer:
The population distributions are normal.
The population variances are equal.

74 If you know that the distribution of family incomes is strongly skewed, will it be appropriate to use analysis of variance to study the effect of level of education on family income?

Answer: No. You cannot assume that the populations are normally distributed.

75 Will it be appropriate to use a *t* test?

Answer: No. The *t* test also assumes normally distributed populations.

76 If comparisons are to be made among several different samples, *t* tests are not appropriate. For example, if you are comparing four sample means, six different two-way comparisons are possible. If you select a significance level of 5%, each individual comparison has a 5% chance of a Type I error. For all six comparisons you have substantially more than a 5% chance of at least one Type I error. The analysis of variance avoids this problem because it makes only one comparison (between s_b^2 and s_w^2). Therefore, when you are making comparisons among a number of samples, the appropriate test to use is (a t test/analysis of variance).

Answer: Analysis of variance

77   The following is a brief summary of the adjustment scores of mental patients treated with three different drugs. Is analysis of variance an appropriate statistical technique to apply in this case?

Group 1	Group 2	Group 3
$\bar{x} = 5$	$\bar{x} = 10$	$\bar{x} = 4$
$s = 4$	$s = 23$	$s = 5$
$n = 10$	$n = 33$	$n = 45$

Answer: No

78   What assumption is violated by the above data?

Answer: The variances are very unequal $s_1^2 = 16$, $s_2^2 = 529$, $s_3^2 = 25$. In this case, you might be able to restructure your data and apply the *chi square* test discussed in chapter 8.

SELF-TEST   If you have successfully completed this chapter you can now use the F distribution to perform statistical tests. You can:

• Test the significance of the difference between two sample variances

• Perform an analysis of variance to determine if the differences among a number of sample means are statistically significant

• Recognize situations for which F tests should not be used because the data do not meet the assumptions of the mathematical model

Now try these review problems. Table I at the back of the book lists any formulas you may need for reference.

1. A number of students are assigned randomly to three classes with three different teaching methods. The following statistics summarize the performance of the three groups on a comprehensive final exam. Can you perform an analysis of variance with these data? What assumptions are involved?

Group 1	Group 2	Group 3
$n = 10$	$n = 11$	$n = 8$
$\bar{x} = 89$	$\bar{x} = 75$	$\bar{x} = 90$
$s^2 = 100$	$s^2 = 81$	$s^2 = 64$
$s = 10$	$s = 9$	$s = 8$

2. The situation is the same as in problem 1 but the data are different. Can you perform an analysis of variance? What assumptions are involved?

Group 1	Group 2	Group 3
$n = 10$	$n = 11$	$n = 8$
$\bar{x} = 90$	$\bar{x} = 86$	$\bar{x} = 70$
$s^2 = 144$	$s^2 = 81$	$s^2 = 64$
$s = 12$	$s = 5$	$s = 4$

3. You believe that a new production procedure will reduce the variability in size of molded plastic parts. You obtain samples of parts molded under the new and old procedures and measure them as summarized below. Outline an appropriate statistical test at the 5% significance level. Is the difference between the variability of the two samples significant? Is the difference between their means significant?

Group 1	Group 2
$n = 15$	$n = 25$
$\bar{x} = 20.00$	$\bar{x} = 21.00$
$s^2 = 0.0625$	$s^2 = 1.00$
$s = 0.25$	$s = 1.00$

4. Perform an analysis of variance of the following data. Use a 5% significance level.

Group 1	Group 2	Group 3
1	2	3
2	3	4
2	4	4
2	4	5
3	5	
	6	

ANSWERS

To review a problem, study the frames indicated after the answer.

1. Yes, an analysis of variance is possible. You must be prepared to assume that the populations from which you are sampling are normally distributed. To interpret the results you must also be prepared to assume that the teaching method is the only reasonable explanation of the differences between groups. *See frames 73 to 78; see also frames 26 to 31 in chapter 4.*

2. No, you cannot perform an analysis of variance. One of the assumptions for analysis of variance is that the populations sampled have equal variances. The variances of these three samples are too different to allow this assumption. The variance of group 1 is nine times as great as that of group 3 and almost six times as great as that of group 2. *See frames 1 to 23 and 73 to 78.*

3.

Null hypothesis	$\sigma_2 = \sigma_1$
Alternative	$\sigma_2 > \sigma_1$
Significance level	$\alpha = 0.05$
Critical region	$F > 2.35$ (df = 24, 14)

$$F = \frac{s_2^2}{s_1^2} = \frac{1.00}{0.0625} = 16.0$$

The difference in variances is significant. With this substantial difference in variances, the accuracy of a t test would be in doubt. *See frames 1 to 23.*

4.

Group 1		Group 2		Group 3	
x	x^2	x	x^2	x	x^2
1	1	2	4	3	9
2	4	3	9	4	16
2	4	4	16	4	16
2	4	4	16	5	25
3	9	5	25		
		6	36		
10	22	24	106	16	66
$n_1 = 5$		$n_2 = 6$		$n_3 = 4$	
$\Sigma x_T = 50$		$\Sigma x_T^2 = 194$		$N = 15$	

Total sum of squares:

$$\Sigma x_T^2 - \frac{(\Sigma x_T)^2}{N} = 194 - \frac{(50)^2}{15} = 194 - \frac{2500}{15} = 194 - 1666.7 = 27.3$$

$$\text{df} = N - 1 = 14$$

Between groups sum of squares:

$$\frac{(\Sigma x_1)^2}{n_1} + \frac{(\Sigma x_2)^2}{n_2} + \frac{(\Sigma x_3)^2}{n_3} - \frac{(\Sigma x_T)^2}{n_1} = \frac{(10)^2}{5} + \frac{(24)^2}{6} + \frac{(16)^2}{4} - 166.7$$

$$= 20 + 96 + 64 - 166.7 = 13.3$$

$$\text{df} = g - 1 = 2$$

Within groups sum of squares:

$$\text{Total} - \text{between} = 27.3 - 13.3 = 14.0$$

$$\text{df} = N - g = 12$$

	Sum of Squares	df	Variance Estimate	F
Total	27.3	14	—	
Between groups	13.3	2	6.65	5.68
Within groups	14.0	12	1.17	

The critical region is $F \geq 3.89$ for a 5% significance level. The difference between groups is significant. *See frames 26 to 69.*

7 The Relation between Two Sets of Measures

So far, we have been concerned with only one kind of data at a time. In this chapter you will learn some ways of studying the relationships between two different measures. For example, you may wonder if the height of a tree is related to its age, or if wealth and happiness go together. A *scattergram* can be used to display graphically the relation between two different measures in a sample. It is also possible to summarize the relation between two measures quantitatively, using a *correlation coefficient*. And, given certain assumptions, it is possible to test the significance of a correlation coefficient by referring to a sampling distribution.

If two measures are related, it is possible to use one measure for an individual to predict the other. This is done by means of a *regression equation*. For example, if college enrollments and purchases of personal computers are related, you can apply a regression equation to enrollment data in a given town to determine the approximate size of the personal computer market in that town.

When you have completed this chapter, you will be able to:

- Construct and interpret a scattergram

- Compute and interpret the correlation coefficient r

- Test the null hypothesis that two measures have zero correlations

- Use a regression equation to predict the value of one measure on the basis of a related measure

SCATTERGRAMS

Often we are interested in the relationship between two different measurements or observations in a population. For example, do high scores on a scholastic aptitude test tend to go with high grade point averages, or are the two measures unrelated? Do rich families tend to have many children or few children, or is there no relation at all between wealth and the number of children in a family? One simple way to examine such a relationship is to draw a scattergram.

1   Look at the scattergram below. Each *x* on the scattergram represents a family in our sample. The height of the *x* on the graph represents the number of children in the family. The left-right position of the *x* represents the family income; for example, A represents a family with three children and an income of $40,000. B represents a family with _____ children and an income of _____ .

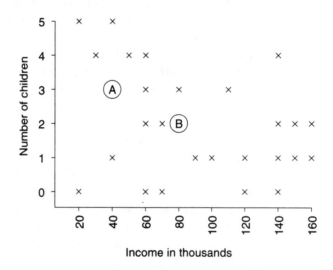

FIGURE 7-1:
Income versus family size

Answer:

Two
$80,000

2   In the scattergram, circle the *x* that represents a family with no children and an income of $20,000.

Answer:

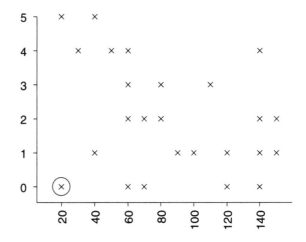

FIGURE 7-2:
$20,000, no children

3 Looking at the scattergram, you can say that in this sample families with high incomes tend to have relatively (larger/smaller) numbers of children.

Answer: Smaller

4 Use figure 7-3 to construct a scattergram for the following data:

Student No.	Midterm Exam Grade	Final Exam Grade
1	65	74
2	68	73
3	71	71
4	75	80
5	75	83
6	85	85
7	88	90
8	95	94
9	97	99
10	100	98

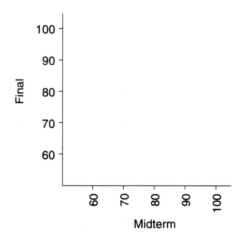

FIGURE 7-3:
Scattergram grid

Answer: Your scattergram should look like the one below.

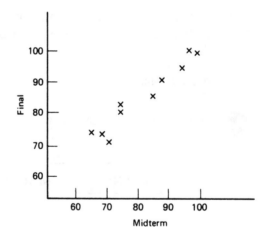

FIGURE 7-4:

Midterm versus
final grades

You might have reversed the axes so that the vertical dimension represented the midterm grade and the horizontal dimension, the final grade.

When one measure may be used to predict another, it is customary to represent the predictor on the horizontal dimension (the x-axis).

5   In this case high scores on the midterm tend to go with (high/low) scores on the final exam.

Answer: High

6 In each of the two scattergrams you have seen, you could make a reasonably good representation of the relationship between the two measurements by drawing a straight line. This kind of relationship is called a linear relationship, as shown below.

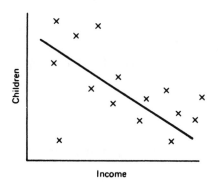
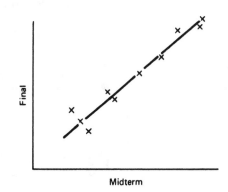

FIGURE 7-5:
A linear relationship

Which of the scattergrams below shows a clear linear relationship?

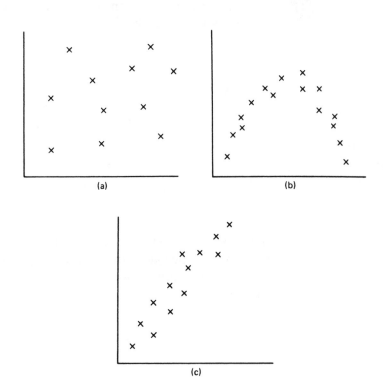

FIGURE 7-6:
Describe the relationship

Answer: (c)

7 Sometimes a relationship between two measurements is curvilinear; that is, it is best described by a curved line. Which of the scattergrams in figure 7-6 represents a curvilinear relationship?

Answer: (b)

8 A relationship that can be described by a straight line is called a _____ relationship.

Answer: Linear

9 A relationship that is best described by a curved line is called a _____ relationship.

Answer: Curvilinear

10 When there is no relationship between two variables, the scattergram will look something like this.

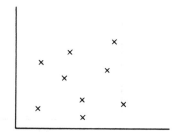

FIGURE 7-7:
Random distribution,
no relationship

The stronger the relationship, the more closely the points on your scattergrarn will approach some linear or curvilinear pattern. Which of the scattergrams below represents the stronger relationship between two variables?

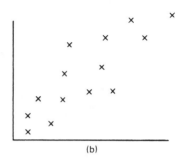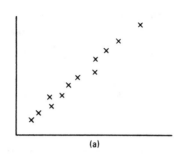

FIGURE 7-8:
Which relationship
is stronger?

(a)   (b)

Answer: (a)

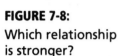 Which of these two scattergrams represents the stronger relationship between two variables?

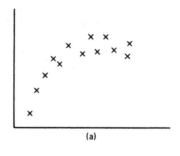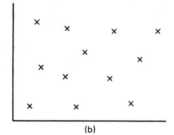

FIGURE 7-9:
Which relationship
is stronger?

(a)   (b)

Answer: (a)

> *In spreadsheet terminology, a scattergram is called an x-y chart. No particular statistical knowledge is required to create a scattergram. Both Microsoft Excel and Lotus 1-2-3 have charting instructions that allow you to select and create a variety of chart types from spreadsheet data. If you are using Microsoft Excel, select Insert|Chart on the main menu. If you are using Lotus 1-2-3, select Tools|Chart. Dialogs will allow you to select x-y as the type of chart and will allow you to select the ranges of data to chart.*

CORRELATION COEFFICIENTS

If two measures have a linear relationship, it is possible to describe how strong the relationship is by means of a statistic called a *correlation coefficient*. The symbol for the correlation coefficient is r. The symbol for the corresponding population parameter is ρ (the Greek letter "rho").

There are various forms of correlation coefficient for specialized types of data. We explain only the basic type, which is called the Pearson product-moment correlation.

12 Match the explanations of correlation coefficients to the appropriate illustrations below.

Correlation Coefficient		Figure
$r = +1.0$	All scores are exactly on the line. High scores on one measure go with high scores on the other.	_____
r between $+1.0$ and 0 (e.g., $+0.5$)	High scores on one measure tend to go with high scores on the other but the relationship is not perfect.	_____
$r = 0$	No relationship at all between the two measures.	_____
r between 0 and -1.0 (e.g., -0.3)	High scores on one measure tend to go with low scores on the other but the relationship is not perfect.	_____
$r = -1.0$	All scores are exactly on the line. High scores on one measure go with low scores on the other.	_____

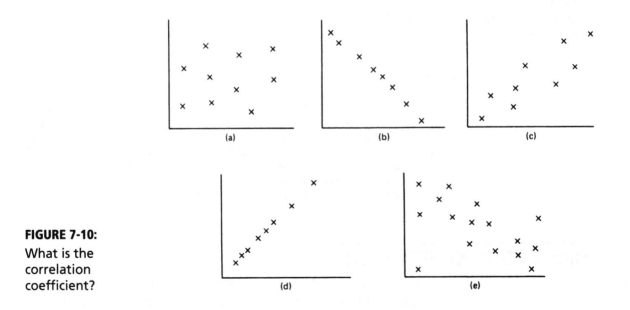

FIGURE 7-10:
What is the correlation coefficient?

Answers:

(d)

(c)

(a)

(e)

(b)

13 The basic formulas for the correlation coefficient are

$$\rho = \frac{1}{N} \Sigma \frac{(x - \mu_x)}{\sigma_x} \cdot \frac{(y - \mu_y)}{\sigma_y}$$

$$r = \frac{1}{n - 1} \Sigma \frac{(x - \bar{x})}{s_x} \cdot \frac{(y - \bar{y})}{s_y}$$

Let us apply this to a simple example. Suppose you are interested in the relation between the age of garden snakes (in months) and their length (in inches). You obtain a sample of three garden snakes of known age and you measure them with the following results:

Snake No.	Age (x)	Length (y)
1	1	4
2	2	6
3	3	8
$\bar{x} = 2$		$\bar{y} = 6$
$s_x = 1$		$s_y = 2$

Since this is a sample, the formula we must use for a correlation coefficient is $r = $ _____ .

Answer:

$$r = \frac{1}{n - 1} \Sigma \frac{(x - \bar{x})}{s_x} \cdot \frac{(y - \bar{y})}{s_y}$$

14 For each snake we must compute $(x - \bar{x})/s_x \cdot (y - \bar{y})/s_y$. Note carefully that $(x - \bar{x})/s_x$ tells us how many standard deviations a given snake's age is away from the mean. It is a form of z score. What does $(y - \bar{y})/s_y$ tell us?

Answer: How many standard deviations the snake's length is away from the mean.

15 Complete this table to find $(x - \bar{x})/s_x \cdot (y - \bar{y})/s_y$ for each snake.

Age x	$(x - \bar{x})$	$\dfrac{x - \bar{x}}{s_x}$	Length y	$(y - \bar{y})$	$\dfrac{y - \bar{y}}{s_y}$	$\dfrac{(x - \bar{x})}{s_x} \cdot \dfrac{(y - \bar{y})}{s_{\bar{y}}}$
1	——	——	4	——	——	——
2	——	——	6	——	——	——
3	——	——	8	——	——	——
	$\bar{x} = 2$			$\bar{y} = 6$		
	$s_x = 1$			$s_y = 2$		

Answer:

Age x	$(x - \bar{x})$	$\dfrac{x - \bar{x}}{s_x}$	Length y	$(y - \bar{y})$	$\dfrac{y - \bar{y}}{s_y}$	$\dfrac{(x - \bar{x})}{s_x} \cdot \dfrac{(y - \bar{y})}{s_{\bar{y}}}$
1	−1	−1	4	−2	−2	+1
2	0	0	6	0	0	0
3	+1	+1	8	+2	+1	+1

16 For this sample what is $\sum (x - \bar{x})/s_x \cdot (y - \bar{y})/s_y$? Add up the sum in your table.

Answer: $1 + 0 + 1 = +2$

17 Compute r.

$$r = \frac{1}{n - 1} \sum \frac{(x - \bar{x})}{s_x} \cdot \frac{(y - \bar{y})}{s_y}$$

Answer:

$$r = \frac{1}{3 - 1} \cdot 2 = +1.0$$

18 A correlation coefficient of +1.0 means that for every pair of observations, $(x - \bar{x})/s_x$ is exactly equal to $(y - \bar{y})/s_y$. When a snake's length is two standard deviations above the mean, its age is _____ .

Answer: Two standard deviations above the mean

19 If the correlation coefficient is +1.0, information about one measurement tells you precisely what the other measurement must be. For example, suppose that instead of three garden snakes, you have a large population. For age the mean is 12 months with a standard deviation of 4 months. For length the mean is 25 inches with a standard deviation of 8 inches. The correlation between age and length is $\rho = +1.0$. You are going to pick one snake at random from this population. Can you predict accurately what its age will be?

Answer: No. Your best guess would be 12 months, but the odds are small that you will be precisely correct.

20 Can you predict accurately what its length will be?

Answer: No

21 You select a snake and determine its age. Can you predict accurately what its length will be?

Answer: Yes. If you know how many standard deviations above or below the mean its age is, you know that its length must be the same number of standard deviations above or below the mean.

22 If a correlation coefficient is −1.0, you can also predict one measure precisely by knowing the other. When, $r = -1.0$, $(x - \bar{x})/s_x = -(y - \bar{y})/s_y$. An x score two standard deviations above the mean must be paired with a y score two standard deviations (above/below) the mean.

Answer: Below

23 You are studying the relationship between the number of lines of code in computer programs and the number of errors ("bugs") found after completion. In a large sample of programs written by computer science students, the correlation between lines of code and number of bugs is +0.96. If you know the number of lines of code in a program, can you state precisely how many bugs will be found in the program?

Answer: No. The relationship is strong but not perfect.

24 On the basis of this correlation coefficient, describe in general terms how lines of code are related to bugs in the sample of programs.

Answer: Longer programs tend to go with larger numbers of bugs.

25 If you know the number of lines of code in a program, can you make a more accurate prediction of the number of bugs that will be found than if the program length were unknown?

Answer: Yes

26 In the case of the garden snakes, we can say that the age of the snakes explains all of the variance in their lengths; that is, knowing their ages, we can predict their lengths precisely. In the case of the computer programs, does the number of lines of code account for all of the variance in the number of bugs?

Answer: No

27 The square of the correlation coefficient tells us exactly what percentage of the variance of y is explained by x. For example, the percentage of the variance in number of bugs explained by differences in program length in our sample of programs is r^2 or _____ .

Answer: 92 percent

28 Another term for r^2 is *covariance*. Ninety-two percent is the _____ of lines of code and number of bugs.

Answer: Covariance

29 If the correlation between aptitude-test scores and grade point averages in a population of students is 0.70, the aptitude-test scores explain _____ percent of the variance in academic performance.

Answer: 49

30 When we say that 49 percent of the variance in academic performance is "explained" by aptitude-test scores, it is important to remember that we are not necessarily describing a cause-and-effect relationship; we are only describing how one variable can be used to predict another. For example, could you use academic performance (grade point averages) to predict aptitude-test scores?

Answer: Yes

31 The correlation between grade point averages and aptitude-test scores is +0.70. Academic performance therefore "explains" $r^2 = 49$ percent of the variance in _____ .

Answer: Aptitude-test scores

32 A researcher is able to determine the total lifetime earnings and the age at which they died for an appropriate sample of working people. The correlation between total lifetime earnings and life span is +0.80. The co-variance of earnings and life span is 0.64. The researcher concludes that poverty is a cause of premature death. Does his statistical information support this conclusion? Why or why not?

Answer: No. The correlation coefficient does not tell which measurement is cause and which effect, or if some other factor underlies both of the measured items. Early death probably reduces total lifetime earnings. Health problems might cause both early death and low earnings. All the researcher can conclude from the correlation is that early death and low lifetime earnings go together.

COMPUTING r

For most practical purposes, correlation coefficients are calculated by computers. Nevertheless, at least once in your life you ought to compute one correlation coefficient by hand so that you can check the honesty of your computer.

33 It is not necessary to compute $(x - \bar{x})/s_x$ and $(y - \bar{y})/s_y$ for every pair of scores to compute r. In practice, less computation is required if you use the following formula:

$$r = \frac{n \sum (xy) - (\sum x)(\sum y)}{\sqrt{\left[n \sum x^2 - (\sum x)^2\right]\left[n \sum y^2 - (\sum y)^2\right]}}$$

Courage! The formula is not as difficult as it looks. You can set up a table similar to those used for computing standard deviations or analysis of variance. Use the information from the following table to complete the computation of r for the paired data $x = 1, 2, 4, 5, 5$, and $y = 2, 4, 5, 7, 8$. (If you have a calculator, this is a good place to use it.)

x	x^2	y	y^2	xy
1	1	2	4	2
2	4	4	16	8
4	16	5	25	20
5	25	7	49	35
5	25	8	64	40
17	71	26	158	105

Answer:

$$r = \frac{5(105) - (17)(26)}{\sqrt{\left[5(71) - (17)^2\right]\left[5(158) - (26)^2\right]}}$$

34 Set up a table and compute r for the following data:

x	y
5	4
6	3
1	2
4	6
2	3

Answer:

x	x^2	y	y^2	xy
5	25	4	16	20
6	36	3	9	18
1	1	2	4	2
4	16	6	36	24
2	4	3	9	6
18	82	18	74	70

Answer:

$$r = \frac{5(70) - (18)\,(18)}{\sqrt{\left[5(82) - (18)^2\right]\left[5(74) - (18)^2\right]}} = \frac{26}{\sqrt{[86][84]}} = +0.41$$

HYPOTHESIS TESTING

As with other statistics, it is possible to draw conclusions about the correlations within a population on the basis of a sample by referring to a sampling distribution. That is, you can test the significance level of a correlation coefficient.

35 If you look at pairs of observations in a small sample, you are quite likely to find correlations by chance. For example, in the space below, write the last seven digits of your social security number in column a and write your telephone number in column b. Now compute the correlation between the two sets of observations. Refer to the formulas in table 1 at the back of the book if you wish, or use

$$r = \frac{n\,\Sigma\,(xy) - (\Sigma\,x)\,(\Sigma\,y)}{\sqrt{\left[n\,\Sigma\,x^2 - (\Sigma\,x)^2\right]\left[n\,\Sigma\,y^2 - (\Sigma\,y)^2\right]}}$$

a (Social Security No.)		b (Telephone No.)		
x	x^2	y	y^2	xy
$\Sigma x =$	$\Sigma x^2 =$	$\Sigma y =$	$\Sigma y^2 =$	$\Sigma xy =$

Answer: You almost certainly obtained some value of *r* other than 0. The probability that it was somewhere between +0.75 and –0.75 is 95%.

36 As you no doubt suspect, we can construct theoretical sampling distributions for *r* under certain assumptions. Specifically, if we assume that we are drawing random samples from a population in which *x* and *y* are both normally distributed and have zero correlation, we can compute the probability of obtaining various values of *r*. This information is presented in table VII at the back of the book. What is the title of the table?

Answer: Critical values of *r*.

37 We can use table VII to test the null hypothesis that the population correlation ρ is zero. For example, a researcher believes that there is a correlation between the blood pressure and the amount of a certain chemical in the blood. She knows that both measures are normally distributed. From a random sample of 27 individuals she computes a correlation coefficient. What is the null hypothesis? $\rho =$ _____ .

Answer: 0

38 What is the alternative?

Answer: $\rho \neq 0$

39 In establishing a critical region, the researcher will consider (both ends/one end) of the sampling distribution.

Answer: Both ends

40 Let us use a 5% significance level. Since both ends of the sampling distribution are being considered, we must find the critical value of *r* that corresponds to 0.025 for our sample size. What is this critical value?

Answer: 0.396

41 What is the critical region?

Answer: $r \geq +0.396$ or $r \leq -0.396$

42   If the researcher obtains a value from her sample of $r = -0.29$, can she reject the null hypothesis?

Answer: No

43   You believe that the speed in words per minute and error rate in errors per page of typists are negatively correlated; that is, fast typists tend to make (more/fewer) errors than slow ones.

Answer: Fewer

44   You will test this theory with a sample of 20 typists. Outline an appropriate statistical test of the theory, using a significance level of 1%.

Null hypothesis   _____

Alternative   _____

Significance level   _____

Critical region   _____

Answer:

Null hypothesis $\rho = 0$
Alternative $\rho < 0$
Significance level $\alpha = 0.01$
Critical region $r \leq -0.516$

45   Using the table of critical values of r involves two assumptions.

(a) The two measures are distributed _____ .

(b) The sample is chosen at _____ .

Answer:

(a) Normally
(b) Random

46   To test your theory about the relation between typing speed and error rate, you choose 5 very fast typists, 5 moderately fast typists, 5 moderately slow typists, and 5 very slow typists for your sample. Does this procedure meet the assumption of your statistical test? Why or why not?

Answer: No. This is not a random sample because not all typists have an equal chance of being selected. You cannot use the table of critical values of *r*, but you can use other statistical procedures, as you will see in a moment.

47 It is not uncommon for researchers doing exploratory studies to collect data on a large number of variables and look for relationships. For example, suppose you collect data from a sample of voters on a large number of variables—age, years of education, income, and so forth. You use a computer to calculate correlation coefficients among 12 variables (66 correlation coefficients). You find that three of these are significant at the 5% level. Do your results support the theory that these three pairs of variables are correlated in the population from which the sample was drawn? Why or why not?

Answer: No. When you calculate 66 correlation coefficients you must expect 5 percent of them (3.3 on the average) to be significant at the 5 percent level by chance.

COMPUTING CORRELATIONS WITH SPREADSHEET PROGRAMS

A word of caution—the names of spreadsheet functions related to the subjects covered in this chapter may be misleading. For example, the spreadsheet functions named CORREL do not return the sample correlation, *r*, as you might expect. Instead, they compute the population parameter ρ.

Nevertheless, you can let a spreadsheet do the work for you. Here are the functions you will need to know:

	Microsoft Excel	**Lotus 1-2-3**
Population correlation	=CORREL(rangex, rangey)	@correl(rangex, rangey)
Population covariance	=COV(rangex, rangey)	@cov(rangex, rangey, 0)
Sample correlation	=PEARSON(rangex, rangey)	Not available; use @sqrt(@cov(rangex, rangey, 1))
Sample covariance	=RSQ(rangex, rangey)	@cov(rangex, rangey, 1)

Notice that for Lotus 1-2-3, @cov type 0 is the *population* covariance and @cov type 1 is the *sample* covariance.

48 You have a theory that the later in the evening you order a pizza, the larger the pizza will be. For two weeks, you order a pizza at some random time during the evening and measure its diameter (before dining). You enter the exact time you placed each pizza order in cells B11 through B24 of a spreadsheet. You enter the diameter in centimeters of each pizza in cells C11 through C24.

What formula will you use to find the sample correlation between time of order and diameter of pizza? (Answer the question for the spreadsheet program you use.) _____

Answer:

If you use Microsoft Excel, =PEARSON(B11:B24, C11:C24)
If you use Lotus 1-2-3, @sqrt(@cov(B11..B24, C11..C24, 1))

The critical values for r presented in table VII at the back of the book are derived from the t distribution. Spreadsheet programs do not provide a separate function for computing the probability of a given value of r or an inverse function that will allow you to find the value of r that corresponds to a given probability level. Instead, you may use the following formula, which relates r to the t distribution:

$$t_{n-2} = \frac{r\sqrt{n-2}}{\sqrt{1-r^2}}$$

The computation is a bit cumbersome to perform by hand, but a spreadsheet performs it easily.

	Microsoft Excel	Lotus 1-2-3
To find the t value	=(r*SQRT(n–2))/SQRT(1–(r*r))	(r*@sqrt(n–2))/@sqrt(1–(r*r))
To find the two-tailed probability	=TDIST(t, n–2, 2)	@tdist(t, n–2, 1, 2)

Continuing with the pizza example, let us assume you have entered formulas to place the sample statistics in the following cells:

r in cell E5

the covariance (r^2) in cell E6

the number of observations (n) in cell E4

49 You want to enter a formula that will display the t value in cell E7 and another formula that will display the significance level of your result in cell E8. (Answer the questions for the spreadsheet program you use.)

What formula will you enter in E7? _____

What formula will you enter in E8? _____

Answer:

If you use Microsoft Excel,
  in E7, =(E5*SQRT(E4–2))/SQRT(1–(E6))
  in E8, =TDIST(E7, E4–2, 2)
If you use Lotus 1-2-3,
  in E7, (E5*@sqrt(E4–2))/@sqrt(1–(E6))
  in E8, @tdist(E7, E5–2, 1, 2)

PREDICTION

Often we wish to use one measure to predict another measure; for example, we may use aptitude tests to predict academic performance or job performance. We may use advertising volume to predict the sales of a product. We may use rainfall to predict the growth rate of crops. The technique for developing such predictions is called *regression analysis*.

50 Consider the scattergram below. It represents the relation between a test of finger dexterity and productivity on an assembly job.

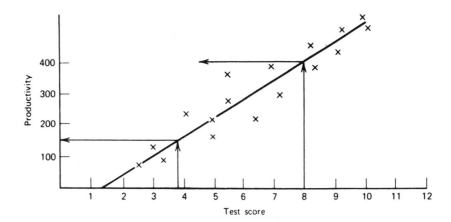

FIGURE 7-11:

Dexterity versus productivity

The diagonal line is the straight line that best represents the relationship between the two measures. You can use this line to obtain the best prediction of productivity for any given test score; for example, if a person scores 3.75 on the test, our best prediction of his productivity is 150 units per day. If a person scores 8.0, our best prediction of his productivity is _____ .

Answer: 400 units per day

51 The line on the scattergram is called the regression line of y on x, or, in this case, the _____ of _____ on _____ .

Answer:

Regression line
Productivity
Test scores

52 The regression line of y on x indicates the best prediction of y for every value of x on the basis of the sample used. If there is no relationship between the two measures, the regression line does not add to your knowledge. If there is no relationship between y and x in your sample (i.e., if $r = 0$), the best prediction of y is always $\bar{y}$, no matter what value x has. For example, try to draw in the regression line of y on x in the scattergram in figure 7-12. You want to draw a line that shows that $\bar{y}$ is the best prediction of y for every value of x.

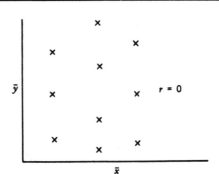

FIGURE 7-12:

Draw the regression line of *y* on *x*

Answer:

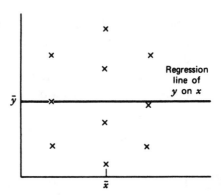

FIGURE 7-13:

The regression line of *y* on *x*

If you use the regression line illustrated above to predict *y*, your prediction will always be $\bar{y}$ no matter what value of *x* you have.

53 When we use a regression line, we assume that the relationship between the two measures is linear. Would a straight regression line give good predictions if the relation between the two measures were strongly curvilinear?

Answer: No

54 It is important to draw a scattergram to make sure that the relationship between the two measures is linear, but it is not necessary to read from the scattergram to make predictions.

The mathematical equivalent of reading the scattergram is accomplished by using the following formula:

$$y = \bar{y} + b(x - \bar{x})$$

where

$$b = r\frac{s_y}{s_x} \text{ or } \frac{n \sum (xy) - (\sum x)(\sum y)}{n \sum x^2 - (\sum x^2)}$$

This formula gives the best possible prediction of y for any given value of x. To use the formula you must find the values of $\bar{x}$, $\bar{y}$, and b on the basis of your sample data and then plug in whatever value of x you require.

Let us apply this procedure. The following data represent advertising expenditures (in thousands of dollars) and sales (in tens of thousands of dollars) for a department store. Your problem is to predict the sales that will result from spending $1,600 on advertising.

Advertising Expense (× $1,000)	Sales (× $10,000)
5	4
6	3
1	2
4	6
2	3

These numbers are the same as those used in frame 34, so you can save yourself some arithmetic by looking back at your computations for that frame. First, what are $\bar{x}$ and $\bar{y}$?

$$\bar{x} = \underline{\hspace{1cm}}$$

$$\bar{y} = \underline{\hspace{1cm}}$$

Answer:

$\bar{x} = 3.6$
$\bar{y} = 3.6$

55 Now compute *b*. Refer to the formula above and to your work on frame 34. You have already computed most of the necessary terms.

$b =$ _____ .

Answer:

$$b = \frac{n \sum (xy) - (\sum x)(\sum y)}{n \sum x^2 - (\sum x)^2} = \frac{5(70) - (18)(18)}{5(82) - (18)^2} = 0.302$$

56 You want to predict the sales resulting from $1,600 of advertising. What value of *x* will you use? (Remember the data are given in thousands of dollars.)

Answer: 1.6

57 Use the regression equation $y = \bar{y} + b(x - \bar{x})$ to predict the value of *y*.

$y =$ _____ .

Answer:

$$y = \bar{y} + b(x - \bar{x}) = 3.6 + 0.302(1.6 - 3.6) = 3.6 - 0.604 = 3.0, \text{ or } \$30,000$$

58 On the basis of the same sample, what sales would you predict for $4,600 of advertising?

Answer:

$$y = \bar{y} + b(x - \bar{x}) = 3.6 + 0.302(4.6 - 3.6) = 3.6 + 0.302 = 3.9, \text{ or } \$39,000$$

59 When you use a regression equation to predict values of a measure, you are subject to the danger that the relationship between the measures in your sample is entirely or partly due to chance. By making certain complex assumptions about the distribution of *y*, it is possible to establish confidence intervals for your predictions. These procedures are beyond the scope of this book. Another useful check on regression equations is cross validation. You can use the regression equation developed with one sample to predict values of *y* for a new sample and then check the accuracy of the predictions. For example, a researcher is developing a method of predicting success in learning foreign languages based on an auditory discrimination test—a listening test. On the basis of his sample, he computes

$$\bar{x} = 10 \qquad\qquad \bar{y} = 50 \qquad\qquad b = 5.0$$

$\quad$ (listening test) $\qquad\qquad$ (language test)

Write the appropriate regression equation for predicting y.

$$y = \underline{\quad\quad} .$$

Answer:

$$y = \bar{y} + b(x - \bar{x}) = 50 + 5.0(x - 10) = 5x$$

60 To cross-validate his regression equation, the researcher will use (a new sample/the same sample).

Answer: A new sample

61 In his new sample the researcher obtains listening test scores of 5, 6, 8, 12, and 15. What language learning scores will he predict? Apply his regression equation.

x	*y* (predicted)
5	_____
6	_____
8	_____
12	_____
15	_____

Answer:

x	*y* (predicted)
5	25
6	30
8	40
12	60
15	75

62   The actual scores he finds are as follows:

x	y (predicted)	y (actual)
5	25	23
6	30	35
8	40	41
12	60	58
15	75	75

His predictions appear to be relatively (accurate/inaccurate).

Answer: Accurate

63   Describe cross-validation briefly in your own words.

Answer: Your answer should have included these points:

(a) Obtain a new sample.

(b) Use regression equation to predict scores.

(c) Compare predicted scores with actual scores.

SELF-TEST   If you have successfully completed this chapter, you can now statistically describe and test the relationship between two measures. You can:

• Draw a scattergram

• Compute the correlation coefficient and test the null hypothesis that $r = 0$

• Use a regression equation to predict one measure on the basis of another, correlated measure

Now try these review problems. Table I at the back of the book lists any formulas you may need for reference.

1.   Draw a scattergram for the following paired data.

   x: 1, 2, 3, 5, 7, 8, 10, 11

   y: 1, 2, 4, 5, 4, 3, 1, 1

2. Describe the relation between x and y in the data for problem 1. Is a correlation coefficient an appropriate measure of the strength of the relationship?

3. Compute r for the following paired data:

   x: 2, 3, 3, 4, 5, 5, 6

   y: 7, 6, 5, 4, 3, 2, 1

4. For the data in problem 3, outline an appropriate statistical test for the null hypothesis that there is no correlation between the populations of x and y scores. Use a 1% significance level. Is the r obtained in problem 3 significant?

5. Use the data from problem 3 to establish a regression equation to predict y. If $x = 5$, what is your best prediction for y?

ANSWERS

To review a problem, study the frames indicated after the answer.

1. Your scattergram should look like this.

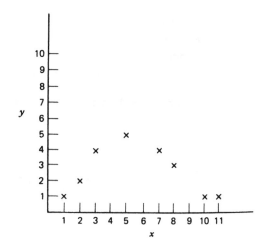

FIGURE 7-14:
Scattergram

See frames 1 to 4.

2. The relationship is curvilinear; therefore r is not an appropriate measure of the strength of the relationship. *See frames 6 to 11.*

3.

x	x^2	y	y^2	xy
2	4	7	49	14
3	9	6	36	18
3	9	5	25	15
4	16	4	16	16
5	25	3	9	15
5	25	2	4	10
6	36	1	1	6
28	124	28	140	94

$$r = \frac{n \sum (xy) - (\sum x)(\sum y)}{\sqrt{\left[n \sum x^2 - (\sum x)^2 \right]\left[n \sum y^2 - (\sum y)^2 \right]}} = \frac{7(94) - (28)(28)}{\sqrt{[7(124) - (784)][7(140) - (784)]}}$$

$$= \frac{-126}{\sqrt{16464}} = \frac{-126}{128} = -0.98$$

See frames 33 to 34.

4.

Null hypothesis	$\rho = 0$
Alternative	$\rho \neq 0$
Significance level	$\alpha = 0.01$
Critical region	$r \geq +0.875$ or $r \leq -0.875$

Since $r = -0.98$, the null hypothesis is rejected. *See frames 35 to 47.*

5.

$$\bar{x} = 4$$

$$\bar{y} = 4$$

$$b = \frac{n \sum (xy) - (\sum x)(\sum y)}{n \sum x^2 - (\sum x)^2} = \frac{7(94) - (28)(28)}{7(124) - (28)^2} = \frac{-126}{84} = -1.5$$

$$y = \bar{y} + b(x - \bar{x}) = 4 - 1.5(x - 4)$$

$$y = 4 - 1.5(5 - 4) = 2.5 \text{ if } x = 5$$

See frames 50 to 63.

8 A Test of Distributions

Many kinds of data come in the form of counts or categories. For example, think of the sets at the start of chapter 1, "blue eyes, brown eyes, green eyes . . . ," or "station wagon, sports car . . . ." Using a sampling distribution called chi square (χ^2), it is possible to determine the probability that any given sample was drawn from a population with a given distribution.

In using chi square for formal hypothesis testing, the important thing is to be able to come up with an appropriate population model for your null hypothesis. With a little ingenuity, it is possible to use chi square tests in many cases for which other kinds of statistical tests are of doubtful applicability, for example, when measurements do not have normal distributions or equal variances.

When you have completed this chapter, you will be able to:

- Perform a χ^2 test with data on a single variable

- Perform χ^2 tests for data on two variables

- Use χ^2 tests as a substitute for analysis of variance or t tests when appropriate

THE CHI SQUARE TEST OF A DISTRIBUTION

The χ^2 (Greek letter chi) distribution is a theoretical sampling distribution that allows you to test the assumption that a sample was drawn from a population with a given distribution. It allows you to compare a sample distribution with a population distribution derived from a theory or null hypothesis and decide whether the sample could reasonably be a random sample from that population.

1 On the basis of genetic theory you predict that a particular population of guinea pigs should be 40 percent brown, 40 percent spotted, and 20 percent white. What distribution of colors would you expect in a sample of 50 guinea pigs? Complete the table below:

	Brown	Spotted	White
Predicted number of guinea pigs			

Answer:

	Brown	Spotted	White
Predicted number of guinea pigs	20	20	10

2 In a sample of 50 guinea pigs you find 30 brown, 15 spotted, and 5 white. This observation is compared with your theoretical prediction in the following table:

	Brown	Spotted	White
Predicted	20	20	10
Observed	30	15	5

Can the differences from what you predicted be accounted for by chance or will you have to reject the theory on which your predictions are based? To find out you can compute a value of χ^2.

The formula for χ^2 is

$$\chi^2 = \Sigma \left[\frac{(f - F)^2}{F} \right]$$

In this formula F is the predicted frequency for a given category (or "cell") and f is the observed frequency. For our example we compute as follows:

f	F	$(f - F)$	$(f - F)^2$	$(f - F)^2 / F$
30	20	+10	100	5
15	20	−5	25	1.25
5	10	−5	25	2.50
			$\Sigma \dfrac{(f - F)^2}{F}$	8.75

What is the value of χ^2 ?

Answer: $\chi^2 = 8.75$

3 If the differences between the theoretically predicted frequencies (F) and the observed frequencies (f) tend to be large, χ^2 will be (large/small).

Answer: **Large**

4 What would a small value of χ^2 indicate?

Answer: **The observed frequencies tend to be close to the theoretical predictions; $(f - F)^2$ tends to be small.**

5 Look at the formula for χ^2. Is a negative value of χ^2 possible?

Answer: **No. $(f - F)^2$ is always positive, so χ^2 is always positive.**

6 To find the probability of obtaining by chance a value of χ^2 as large as 8.75, you can refer to table VIII at the back of the book. Look at the table now. What additional information do you need to use the table?

Answer: You need to know the degrees of freedom, df.

7 In this case the degrees of freedom for χ^2 are the number of categories minus 1. What degrees of freedom will you use for the guinea pig problem?

Answer: 2

8 What is the probability of a χ^2 as large as 8.75?

Answer: Less than 0.025 (between 0.025 and 0.01)

9 You believe that people are most likely to attempt suicide on a weekend. To test this theory, you have obtained reports of attempted suicides from the police. In the past two years, 147 attempts were reported, distributed as follows:

	Sun.	Mon.	Tues.	Wed.	Thur.	Fri.	Sat.	Total
Number of attempts	32	10	13	13	4	40	35	147

An appropriate null hypothesis for a statistical test would be that equal numbers of attempts are made on all days of the week. Use this theory to compute predicted frequencies for the following table. Divide the total equally among the groups.

	Sun.	Mon.	Tues.	Wed.	Thurs.	Fri.	Sat.	Total
Predicted								

Answer:

	Sun.	Mon.	Tues.	Wed.	Thur.	Fri.	Sat.	Total
Predicted	21	21	21	21	21	21	21	147

(Since there are seven categories, the predicted number for each category would be 1/7 of the total.)

10 What are the degrees of freedom for χ^2 in this case?

Answer: df = 6

11 Use table VIII to establish a critical region for χ^2 at a 1% significance level.

Answer: $\chi^2 \geq 16.81$

12 Compute χ^2. Can you reject the null hypothesis?

Answer:

f	F	(f − F)	(f − F)^2	(f − F)^2 / F
32	21	11	121	5.8
10	21	−11	121	5.8
13	21	−8	64	3.0
13	21	−8	64	3.0
4	21	−17	289	13.8
40	21	19	361	17.2
35	21	14	196	9.3
				57.9

$\chi^2 = 57.9$
The null hypothesis can be rejected.

13 There is one important fact to note about this statistical test. The order of the categories has no effect on the value of χ^2. Only the amount of the differences matters. As a result, you would have obtained the same value of χ^2 if your distribution had been the following:

	Sun.	Mon.	Tues.	Wed.	Thur.	Fri.	Sat.	Total
Number of attempts	13	13	35	40	32	10	4	147

Would these results have supported your theory? Why or why not?

Answer: No. These data indicate that suicide attempts are made in midweek.

14 When you use a χ^2 test to reject a null hypothesis, you must always look back at the data to make sure that they support your alternative. Suppose you have taken a census of the insect population of your rose garden. On the basis of several large samples you conclude that the insect population is distributed as follows:

Ladybugs	20%
Brown spiders	20%
Weevils	30%
Aphids	25%
Inchworms	5%

Now you treat your garden with an insecticide that is supposed to control the undesirable weevils, aphids, and inchworms without affecting ladybugs or brown spiders.

To check the effect of the insecticide, you collect 150 insects at random. Your sample is composed as follows:

Ladybugs	25
Brown spiders	45
Weevils	45
Aphids	25
Inchworms	10

Use the census to determine the predicted frequencies. Compute χ^2 and answer these two questions:

(a) Has the distribution of insects changed significantly (at the 5% level)?

(b) Has the insecticide had the intended effect?

Answer:

f	F	(f – F)	(f – F)²	(f – F)² / F
25	30	5	25	0.83
45	30	15	225	7.50
45	45	0	0	0
25	37.5	12.5	156.25	4.17
10	7.5	2.5	6.25	0.83
				13.33

$\chi^2 = 13.33$
df = 4
Critical region: $\chi^2 \geq 9.49$

(a) The distribution of insects has changed significantly.

(b) The insecticide does not have the intended effect. In theory the proportions of ladybugs and brown spiders should increase as the proportions of weevils, aphids, and inchworms decrease. Instead, weevils were unaffected, ladybugs decreased, and inchworms increased. Only brown spiders and aphids changed in the predicted direction.

CHI SQUARE TEST WITH TWO VARIABLES

Chi square can also be used to test hypotheses about distributions based on two variables. To do so we must establish a set of categories based on the two variables and predict the frequency of observations in each category.

15 A researcher believes that people tend to choose mates with the same hair color. She can test this theory by constructing a distribution based on two variables: the husband's hair color and the wife's hair color. Her table will look like this:

Wife	Husband			
	Red	Blond	Black	Brown
Red				
Blonde				
Black				
Brown				

How many cells or categories are in this table?

Answer: 16. Each possible combination of husband and wife is a category.

16 The researcher's first problem is to calculate a predicted frequency for each cell. To do this she must consider the numbers of men and women in each category. She has observed, let us say, 500 couples. The numbers of men and women in each category are indicated in the following table. In addition, the percentage of men in each category is listed at the bottom for reasons you will understand shortly.

Wife	Husband				Total	%
	Red	Blond	Black	Brown		
Red					50	
Blonde					150	
Black					150	
Brown					150	
Total	50	100	150	200	500	
%	10%	20%	30%	40%	100%	

The totals along the edges of the table are called marginals. Look at the marginals to find out how many of the wives are blondes.

Answer: 150

17 How many of the husbands have brown hair? What percent?

Answer:

200
40%

18 Forty percent of all the husbands have brown hair. If there is no relation between the husband's hair color and the wife's hair color, approximately 40 percent of the women in each category should have husbands with brown hair. For example, of the 150 blondes, how many should have husbands with brown hair?

Answer: $0.40 \times 150 = 60$

19 Ten percent of all husbands have red hair. How many of the 50 red-headed wives should have red-headed husbands?

Answer: $0.10 \times 50 = 5$

20 How many of the brown-haired wives should have blond husbands?

Answer: $0.20 \times 150 = 30$

21 Apply this method to fill in predicted frequencies in the table.

Wife	Husband				Total	%
	Red	Blond	Black	Brown		
Red	$F =$	$F =$	$F =$	$F =$	50	10%
Blonde	$F =$	$F =$	$F =$	$F =$	150	30%
Black	$F =$	$F =$	$F =$	$F =$	150	30%
Brown	$F =$	$F =$	$F =$	$F =$	150	30%
Total	50	100	150	200	500	100%
%	10%	20%	30%	40%	100%	

Answer:

		Husband				
Wife	Red	Blond	Black	Brown	Total	%
Red	F = 5	F = 10	F = 15	F = 20	50	10%
Blonde	F =15	F = 30	F = 45	F = 60	150	30%
Black	F =15	F = 30	F = 45	F = 60	150	30%
Brown	F =15	F = 30	F = 45	F = 60	150	30%
Total	50	100	150	200	500	100%
%	10%	20%	30%	40%	100%	

22 Degrees of freedom in this case depend on the number of categories in each direction. They are $(c - 1)(r - 1)$, the number of columns minus one times the number of rows minus one. Columns are the categories of husbands. How many columns are there in this problem?

Answer: 4

23 The rows in this problem are the categories for _____ .

Answer: Wives

24 In this problem df = $(c - 1)(r - 1)$ = _____ .

Answer: $(4 - 1)(4 - 1) = 9$

25 We have added the observed frequencies, f, to the table. Compute χ^2 and answer two questions.

(a) Is χ^2 significant at the 1% level?

(b) Do the data support the researcher's theory that people tend to select mates with the same hair color?

Wife	Husband				Total	%
	Red	**Blond**	**Black**	**Brown**		
Red	F = 5 f = 10	F = 10 f = 10	F = 15 f = 10	F = 20 f = 20	50	10%
Blonde	F = 15 f = 10	F = 30 f = 40	F = 45 f = 50	F = 60 f = 50	150	30%
Black	F = 15 f = 13	F = 30 f = 25	F = 45 f = 60	F = 60 f = 52	150	30%
Brown	F = 15 f = 17	F = 30 f = 25	F = 45 f = 30	F = 60 f = 78	150	30%
Total	50	100	150	200	500	100%
%	10%	20%	30%	40%	100%	

Answer:

f	F	(f – F)	$(f - F)^2$	$(f - F)^2 / F$
10	5	5	25	5.00
10	15	– 5	25	1.67
13	15	– 2	4	0.27
17	15	2	4	0.27
10	10	0	0	0
40	30	10	100	3.33
25	30	– 5	25	0.83
25	30	– 5	25	0.83
10	15	– 5	25	1.67
50	45	5	25	0.56
60	45	15	225	5.00
30	45	– 15	225	5.00
20	20	0	0	0
50	60	– 10	100	1.67
52	60	– 8	64	1.07
78	60	18	324	5.40
				$\chi^2 = 32.57$

(a) χ^2 is significant; df = 9. The critical region is $\chi^2 \geq 21.67$.

(b) The data support the theory because the observed frequency of matched husband and wife couples is greater than predicted in all cases, whereas most of the other observed frequencies are equal to or less than the prediction.

26 As a rule, to use the χ^2 test the predicted frequency for each cell should be at least 5.

There are exceptions to this rule but they are beyond the scope of this book.

Could you have performed the χ^2 test in frame 18 if you had a sample of 100 couples?

Answer: No. Many of the predicted frequencies would be less than 5.

27 You are investigating the sensitivity of goldfish to industrial pollutants in their water. Some fish appear to be very sensitive. They turn white and die instantly. Others are moderately sensitive. They show signs of distress, but do not die. Many show no reaction at all. Since you are working with both black and red goldfish, you would like to know if there is evidence of a difference in sensitivity for the two colors. The following table classifies a group of goldfish according to color and sensitivity. Use the marginals to calculate predicted frequencies on the assumption that there is no relation between color and sensitivity and write these frequencies in the table.

Color		Sensitivity High	Medium	Low	Total	%
Red	F					
	f	4	4	22	30	60%
Black	F					
	f	6	6	8	20	40%
Total		10	10	30	50	

Answer:

Color		High	Medium	Low	Total	%
		Sensitivity				
Red	F	6	6	18		
	f	4	4	22	30	60%
Black	F	4	4	12		
	f	6	6	8	20	40%
Total		10	10	30	50	

28 Can you use the χ^2 test on these data as they are? Why or why not?

Answer: No. Two of the predicted frequencies are less than 5.

29 When the predicted frequencies are less than 5, it is sometimes possible to combine categories so that the predicted frequencies are large enough. In this case you could logically combine two of the sensitivity categories in order to obtain predicted frequencies larger than 5. Prepare a new table that does this. Your table will now have only four cells.

Answer:

Color		High or Moderate	Low	Total	%
		Sensitivity			
Red	F	12	18		
	f	8	22	30	60%
Black	F	8	12		
	f	12	8	20	40%
Total		20	30	50	

30 What are the degrees of freedom for this new table?

Answer: (2 − 1) (2 − 1) = 1

USING A SPREADSHEET TO LOOK UP
CRITICAL VALUES OF χ^2

As you would expect, there are spreadsheet functions that return the probability of χ^2 equal to or greater than a given value. And there are inverse functions that return the critical value of χ^2 that corresponds to a given significance level.

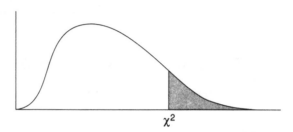

FIGURE 8-1:

The χ^2 distribution

The χ^2 distribution is similar to the F distribution in that it is not symmetrical; its value can never be negative. The critical region consists of large positive values of χ^2.

Look at Figure 8-1. The shaded area under the curve is the probability referred to by the spreadsheet functions. The spreadsheet programs return a probability appropriate for a one-tailed test in the same way that most printed tables do.

Since the χ^2 distribution depends on degrees of freedom, you must always specify degrees of freedom when you use the spreadsheet functions. For Microsoft Excel, there are two different functions, CHIDIST and CHIINV. For Lotus 1-2-3, there is one function name, but there are two types: type 0 returns a probability and type 1 returns a critical value of χ^2.

	Microsoft Excel	Lotus 1-2-3
To find a probability	=CHIDIST(X^2, df)	@chidist(X^2, df, 0)
To find a critical value of χ^2	=CHIINV(p, df)	@chidist(p, df, 1)

31 Write the formula you would enter into a cell of your spreadsheet to have your program calculate the χ^2 score that corresponds to a 5% significance level with 9 degrees of freedom. (Answer the question for the spreadsheet program you use.) _____

Answer:

If you use Microsoft Excel, = CHIINV(0.05, 9)
If you use Lotus 1-2-3, @chidist(0.05, 9,1)

32 Write the formula you would enter into a cell of your spreadsheet to have your program calculate the probability of a χ^2 score of 11.0705 with 5 degrees of freedom. (Answer the question for the spreadsheet program you use.) _____

Answer:

If you use Microsoft Excel, =CHIDIST(11.0705, 5)
If you use Lotus 1-2-3, @chidist(3.9174, 5,0)

Both Microsoft Excel and Lotus 1-2-3 include χ^2 test functions, which perform a χ^2 test to compare actual data to predicted data. These functions require you to identify the two ranges containing the data you want to compare; they return a probability value in the cell you have selected. Note, however, that when you use a computer to calculate chi square you must still decide on the predicted frequencies and compute them by hand.

When you use the χ^2 test functions, you identify the ranges that contain the two sets of sample data. The spreadsheet program then computes the one-tailed probability of the sample result.

The formulas are:

Microsoft Excel	**Lotus 1-2-3**
=CHITEST(range1, range2)	@chitest(range1, range2)

33 For example, suppose your observed data are in cells A10 through A25 and your predicted data are in cells B10 through B25. What formula will you use to test the hypothesis that the observed data are due to random variation? (Answer the question for the spreadsheet program you use.)

_____

Answer:

If you use Microsoft Excel, =CHITEST(A10:A25, B10:B25)
If you use Lotus 1-2-3, @chitest(A10..A25, B10..B25)

WHEN TO USE CHI SQUARE

The *t* test for comparing two means and the analysis of variance both involve assumptions that the populations are normally distributed and have equal variances. A χ^2 test can often be used when the data do not meet these assumptions. Instead of using the measurements to calculate a mean, you simply use them to categorize individual observations, for example, as "high," "medium," and "low," or above and below the mean. The choice of categories should be guided by your theory and by the possibility of constructing an appropriate set of theoretical frequencies.

34 Here is a problem you considered earlier. The following is a summary of the adjustment scores of mental patients treated with three different drugs.

Group 1	Group 2	Group 3
$\bar{x} = 5$	$\bar{x} = 10$	$\bar{x} = 4$
$s^2 = 4$	$s^2 = 23$	$s^2 = 5$
$n = 10$	$n = 33$	$n = 45$

Analysis of variance is not appropriate because the variances of the three groups are unequal. What categories could you use to set up a χ^2 test?

Answer: Group 1, group 2, group 3, high adjustment score, and low adjustment score.

35 When you can use a *t* test or analysis of variance, it is better to do so because the power of these tests is greater than the power of a χ^2 test; that is, you are more likely to be able to reject the null hypothesis if your theory is true. You are interested in whether educational background makes a difference in scores on a troubleshooting test. You have three samples of 25 people each with the following backgrounds: college, high school, and technical institute. The troubleshooting scores appear to be normally distributed, and the sample variances are all quite similar. Could you use χ^2 for examining these data?

Answer: Yes, by using the troubleshooting scores to categorize individuals, for example, as high, medium, and low.

36 Could you use analysis of variance for examining these data?

Answer: Yes

37 Which type of statistical test would you normally choose in this case?

Answer: Analysis of variance

38 You want a quick, informal check on approximately how unusual a distribution is. Either χ^2 or analysis of variance could be used. Which one will you choose?

Answer: We would choose χ^2 because it is usually easier to compute, and precise formal hypothesis testing is not required in this case.

SELF-TEST

If you have successfully completed this chapter, you can now use the chi square distribution to test hypotheses about the distribution of a population. You can:

- Develop an appropriate theoretical population distribution and perform a χ^2 test with frequency data on a single variable

- Develop an appropriate theoretical population distribution and perform a χ^2 test with frequency data on two variables

- Recognize situations in which χ^2 can be used as a substitute for analysis of variance or t tests

Now try these review problems. Table I at the back of the book lists any formulas you may need for reference.

1. A factory has four machines that produce molded parts. A sample of 500 parts is collected for each machine, and the number of defective parts in each sample is determined:

	Machine			
	1	2	3	4
Defects per 500	10	25	0	5

Is there a difference between the machines? Outline an appropriate statistical test and, if possible, determine if these data are significant.

2. The situation is the same as in problem 1, but you are measuring the mean diameter of the parts. You suspect that the mean diameters for parts produced on the four machines are not the same. What would be your first choice of a statistical test? What assumptions would be required?

3. The dropout rate among volunteer workers in community development programs varies widely. The theory is advanced that the degree of involvement of volunteers in setting the goals of the program influences the dropout rate. In a new program, 27 volunteers are selected to participate in a series of goalsetting workshops as part of their activities, whereas 23 others are simply assigned tasks. At the end of two months, the results are as follows:

	Remained in Program	Dropped Out	Total
Workshop group	18	9	27
No workshop group	10	13	23
Total	28	22	

Are these results significant? Outline an appropriate statistical test and compute the necessary statistic.

4. As part of an attitude survey, a sample of men and women is asked to rate a number of statements on a scale of 1 to 5, according to whether they agree or disagree. The following are the results for one of the statements.

	Agree Strongly				Disagree Strongly	
	1	2	3	4	5	Total
Women	3	13	10	16	7	50
Men	2	12	26	10	1	50
Total	5	25	36	26	8	100

Use χ^2 to determine whether there is a significant difference between the answers of men and women.

To review a problem, study the frames indicated after the answer.

1. You can use χ^2 for this problem. The theoretical distribution would call for an equal number of defects from each machine.

	Machine				
	1	2	3	4	Total
F	10	10	10	10	40
f	10	25	0	5	40

F is greater than 5 for all cells.

f	F	$(f - F)$	$(f - F)^2$	$(f - F)^2 / F$
10	10	0	0	0
25	10	15	225	22.5
0	10	− 10	100	10.0
5	10	− 5	25	2.5

$$\chi^2 = 35.0$$
$$df = 3$$

For $\alpha = 0.05$ the critical region is $\chi^2 \geq 7.81$.

For $\alpha = 0.01$ the critical region is $\chi^2 \geq 11.34$.

There is a significant difference among machines. *See frames 1 to 11.*

2. Your first choice should be analysis of variance, since you are dealing with measurements rather than count data. This test will give you the greatest power and the best chance of rejecting the null hypothesis if your theory is true. To use analysis of variance, you must

assume that the populations of measurements for all machines are approximately normally distributed and have approximately equal variances. *See frames 34 to 38; see also frames 73 to 78 in chapter 6.*

3. Chi square is an appropriate test. The null hypothesis is that the workshop has no effect on the dropout rate and that equal numbers drop out from both groups. The alternative is that different numbers will drop out. The critical region for $\alpha = 0.05$ is $\chi^2 \geq 3.84$ (df = 1).

f	F	$(f - F)$	$(f - F)^2$	$(f - F)^2 / F$
18	15.21	2.88	8.29	0.548
10	12.88	− 2.88	8.29	0.644
9	11.88	− 2.88	8.29	0.698
13	10.12	2.88	8.29	0.819
				$\chi^2 = 2.709$

The results are not significant. *See Frames 12 to 14.*

4. Half of the sample is women and half men. If there is no difference in their responses, you would expect equal numbers of men and women in each category of response. The predicted frequencies on this basis are

	1	2	3	4	5
Women	2.5	12.5	18	13	4
Men	2.5	12.5	18	13	4

Because four of the predicted frequencies are less than 5, you must combine categories. By combining categories 1 and 2, and 4 and 5, you may compute χ^2.

		1 and 2	3	4 and 5
Women	F	15	18	17
	f	16	10	23
Men	F	15	18	17
	f	14	26	11

f	F	(f – F)	(f – F)²	(f – F)² / F
16	15	1	1	0.067
14	15	−1	1	0.067
10	18	−8	64	3.556
26	18	8	64	3.556
23	17	6	36	2.118
11	17	−6	36	2.118
				$\chi^2 = 11.482$

For $\alpha = 0.01$ the critical region is $\chi^2 \geq 9.210$ (df = 2). The difference is significant. *See frames 26 to 30.*

9 The Combined Effects of Two Variables

Analysis of variance can be extended to study the combined effect of two experimental treatments on a measure. This method of analysis allows for the design of experiments that are particularly efficient. That is, you can test a large number of hypotheses from relatively few experimental observations. Despite what you may feel when you finish this book, experimental observations are usually more work than statistical analysis. As a result, two-way analysis of variance is a very popular technique for experimenters.

When you have completed this chapter, you will be able to perform and interpret a simple two-way analysis of variance.

1 Often a researcher is interested in the combined effects of two variables on a third measure. For example, consider this problem. A researcher is trying to determine the conditions that will produce the greatest yield from melon vines. He suspects that both the amount of fertilizer used and the amount of water the plants receive will influence the number of melons on a vine. In this case, the two variables whose effect is under study are _____ and _____ .

Answer:

Amount of fertilizer
Amount of water

2   The measure under study is _____ .

Answer: Number of melons per vine

3   To study the effect of the amount of water alone, the researcher might cultivate 20 melon vines with heavy watering and 20 with light watering and count the number of melons per vine in each group. What sort of statistical test could he use? What assumptions would be involved?

Answer: He could use a *t* test for the difference between two means, assuming that the number of melons per vine is approximately normally distributed and that the variances would be approximately equal under the two conditions. If he could not make these assumptions, he could use a χ^2 test by counting the number of vines under each condition that produced above or below average numbers of melons.

4   To study the effect of amount of fertilizer alone, he might cultivate three groups of vines, one with low fertilizer, one with a moderate amount, and one with a high amount. What sort of statistical test could he use for this experiment? What assumptions are involved?

Answer: He could use analysis of variance, assuming approximately normal distributions and equal variances, or he could use a χ^2 test.

5   Suppose that there is an interaction between the effects of the water and fertilizer. Perhaps fertilizer is effective only if there is also heavy watering or perhaps fertilizer helps to compensate for the effects of light watering. Will a study of watering alone or fertilizing alone detect this interaction?

Answer: No

6   In this kind of situation it is possible, in effect, to perform both studies at once and study the results with a two-way analysis of variance. To do this the experimenter will set up six experimental groups as follows:

	Fertilizer		
Water	Light	Medium	Heavy
Heavy			
Light			

If he selects 10 plants at random for each group, how many plants will he have in all?

Answer: 60

7 In all, how many plants will receive heavy watering?

Answer: 30

8 In all, how many plants will receive light fertilizing?

Answer: 20

9 Let us establish some terminology. We call each group of ten a cell; for example, the plants with heavy watering and medium fertilizer make up a cell. We call each level of watering a row; for example, all plants with light watering make up a row. We call each level of fertilizing a column; for example, all plants with light fertilizing make up a column.
All plants with medium fertilizing would be called a _____ .

Answer: Column

10 All plants with medium fertilizing and light watering are called a _____ .

Answer: Cell

11 All plants with heavy watering are called a _____ .

Answer: Row

12 Here are the results of the experiment:

| | Fertilizer | | | | | | | |
| | Light | | Medium | | Heavy | | Totals | |
Water	Number of Melons	χ^2	Number of Melons	χ^2	Number of Melons	χ^2	χ	χ^2
Heavy	1	1	4	16	6	36		
	2	4	5	25	7	49		
	2	4	5	25	7	49		
	3	9	6	36	8	64		
	3	9	6	36	8	64		
	3	9	6	36	8	64		
	3	9	6	36	8	64		
	4	16	7	49	9	81		
	4	16	7	49	9	81		
	5	25	8	64	10	100		
	30	102	60	372	80	652	170	1126
Light	5	25	3	9	0	0		
	6	36	4	16	1	1		
	6	36	4	16	1	1		
	7	49	5	25	2	4		
	7	49	5	25	2	4		
	7	49	5	25	2	4		
	7	49	5	25	2	4		
	8	64	6	36	3	9		
	8	64	6	36	3	9		
	9	81	7	49	4	16		
	70	502	50	262	20	52	140	816
Totals	100	604	110	634	100	704		
Grand total							310	1942

As you can see, we have already computed Σx and Σx^2 for each cell, row, and column and for the total group. What is Σx for the heavy row?

Answer: 170

13 What is Σx^2 for the light column?

Answer: 604

14 What is $\Sigma\, x_T$?

Answer: 310

15 As you know, analysis of variance allows us to estimate the variance of the population on the basis of the differences between group means and to compare this variance estimate with an estimate based on the individual differences within groups. We can group the data for a variance estimate based on group means in several ways. We can look at six individual cell means, at two row means, or at three column means. How many different variance estimates can we obtain from these groupings?

Answer: Three. The variance of the individual cell means can be used to estimate population variance; so can the variance of row means and the variance of column means.

16 Here are the formulas you will use for a two-way analysis of variance. Some of them you have already used for the one-way analysis of variance. Some of them are new. Which ones are new? Place a check mark next to the new ones.

Total sum of squares:

$$\Sigma\, x_T^2 - \frac{(\Sigma\, x_T^2)}{n}$$

$$df = N - 1$$

Between-groups sum of squares:

$$\frac{(\Sigma\, x_1)^2}{n} + \frac{(\Sigma\, x_2)^2}{n} + \cdots - \frac{(\Sigma\, x_T)^2}{N}$$

$$df = rc - 1$$

Between-rows sum of squares:

$$\frac{(\Sigma\, x_{row\,1})^2}{nc} + \frac{(\Sigma\, x_{row\,2})^2}{nc} + \cdots - \frac{(\Sigma\, x_T)^2}{N}$$

$$df = r - 1$$

Between-columns sum of squares:

$$\frac{\left(\sum x_{col1}\right)^2}{nr} + \frac{\left(\sum x_{col2}\right)^2}{nr} + \cdots - \frac{\left(\sum x_T\right)^2}{N}$$

$$df = c - 1$$

Interaction sum of squares:

Between-groups sum of squares − (between-rows sum of squares + between-columns sum of squares), or

$$\frac{\left(\sum x_1\right)^2}{n} + \frac{\left(\sum x_2\right)^2}{n} + \cdots - \frac{\left(\sum x_{row1}\right)^2}{nc} + \frac{\left(\sum x_{row2}\right)^2}{nc} + \cdots - \frac{\left(\sum x_{col1}\right)^2}{nr} + \frac{\left(\sum x_{col2}\right)^2}{nr} + \cdots - \frac{\left(\sum x_T\right)^2}{N}$$

$$df = (r - 1)(c - 1)$$

Within-groups sum of squares:

Total sum of squares − between-groups sum of squares, or

$$\left[\sum x_1^2 - \frac{\left(\sum x_1^2\right)}{n}\right] + \left[\sum x_2^2 - \frac{\left(\sum x_2^2\right)}{n}\right] + \cdots$$

$$df = N - rc$$

Answer: The new ones are:

Between-rows sum of squares
Between-columns sum of squares
Interaction sum of squares

17 The between-groups sum of squares is based on the cells. It tells you exactly the same thing it would in a one-way analysis of variance. If you divide the between-groups variance by the within-groups variance and obtain a significant value of F, what does it tell you? Think of the null hypothesis.

Answer: It tells you that all the cell means are not equal.

18 Will a significant between-groups variance tell you anything about whether water or fertilizer, or both, is the cause of the differences?

Answer: No. You know only that at least one cell mean is significantly different from the others.

19 Look closely at the formula for between-groups sum of squares. It is identical to the one you used for one-way analysis of variance. In the formula for degrees of freedom, r stands for the number of rows and c for the number of columns. In our problem what are the degrees of freedom for the between-groups variance?

Answer: $(2 \times 3) - 1 = 5$

20 Now look at the formulas for between-rows sum of squares and between-columns sum of squares. They are the same as the between-groups formula, except that they are based on rows and columns instead of the individual cells. In our problem, what value corresponds to Σx_{row1}?

Answer: 170

21 What value corresponds to Σx_{col1}?

Answer: 100

22 In the formula for between-rows sum of squares, n stands for the number of measurements in a cell and c for the number of columns. The number of measurements in a row is nc. In our problem this is _____ .

Answer: 30

23 If we divide the between-rows variance estimate by the within-groups variance estimate and obtain a significant F, we can conclude that _____ makes a difference in the mean number of melons on a plant.

Answer: Water

24 If the between-columns variance estimate is significant, we can conclude that _____ makes a difference.

Answer: Fertilizer

25 The between-groups sum of squares will usually be larger than the sum of the between-rows and between-columns sums of squares; that is, part of the variance between cells will not be explained by either the difference between rows or the difference between columns. This remaining part of the difference between cells must be due to the combined effect of the two variables: that is, interaction.

Suppose that fertilizer improves the yield of vines under most conditions, but heavy watering combined with a medium application of fertilizer washes away the fertilizer so that it has no effect. In this case you would expect two significant variance estimates: _____ and _____ .

Answer:

Between columns
Interaction

26   The results of a two-way analysis of variance for the melon problem are summarized below:

	Sum of Squares	df	Variance Estimate	F
Total	340.33	59		
Between rows	15.00	1	15.00	11.28
Between columns	3.33	2	1.67	1.26
Interaction	250.00	2	125.00	93.99
Within groups	72.00	54	1.33	

(The between-groups sum of squares is 268.33, but this figure is not usually included in a summary because the rows, columns, and interaction sums of squares give all the information.)

Are any of these F ratios significant? If so, which ones?

Answer: Between rows and interaction are both significant. Note that the F table has no entry corresponding to df = 54. The values for df = 50 and df = 55 are close enough to ensure that the results are significant.

27   To interpret the results of a two-way analysis of variance, it is helpful to examine the cell means and compare them with a mean for each row and column and a grand total mean.

| | Fertilizer | | | |
Water	Light	Medium	Heavy	All Conditions
Heavy	3.0	6.0	8.0	5.7
Light	7.0	5.0	2.0	4.7
All conditions	5.0	5.5	5.0	5.2 — Grand total

What combination produces the greatest yields?

Answer: Heavy watering and heavy fertilizer

28 What combination produces the poorest yields?

Answer: Light watering and heavy fertilizer

29 You are planning to raise melons in a situation in which you cannot control or predict the amount of watering. There is no irrigation, and rainfall is unpredictably light or heavy. A low yield would be a disaster. What level of fertilizer would you use on the basis of the data in the experiment?

Answer: Medium. It ensures a relatively high yield no matter what the level of watering. Either heavy or light carries a risk of lower yields, although both also bring a chance of higher yields.

ANALYSIS OF VARIANCE BY COMPUTER

Of course, the labor of computing an analysis of variance can be handled by a computer with great ease. You can set up a spreadsheet for the various steps involved in performing an analysis of variance by following the models outlined for hand computation.

Microsoft Excel provides a packaged model under the menu choice for Tools | Data Analysis that happens to correspond to the melon problem. Using the packaged model means that you do not even have to enter the formulas. However, it also means that the formulas are not available in your spreadsheet so that the analysis does not update itself as you change data or assumptions.

A printout of the results of the melon problem, using Microsoft Excel's ANOVA: Two-Factor With Replication, looks like this:

SUMMARY	Lots	Medium	Little	Total
Heavy				
Count	10.00	10.00	10.00	30.00
Sum	30.00	60.00	80.00	170.00
Average	3.00	6.00	8.00	5.67
Variance	1.33	1.33	1.33	5.61
Light				
Count	10.00	10.00	10.00	30.00
Sum	70.00	50.00	20.00	140.00
Average	7.00	5.00	2.00	4.67
Variance	1.33	1.33	1.33	5.61
Total				
Count	20.00	20.00	20.00	
Sum	100.00	110.00	100.00	
Average	5.00	5.50	5.00	
Variance	5.47	1.53	10.74	

ANOVA

Source of Variation	SS	df	MS	F	P-value	F crit
Sample	15.00	1.00	15.00	11.25	0.00	4.02
Columns	3.33	2.00	1.67	1.25	0.29	3.17
Interaction	250.00	2.00	125.00	93.75	0.00	3.17
Within	72.00	54.00	1.33			
Total	340.33	59.00				

WHEN TO USE TWO-WAY ANALYSIS OF VARIANCE

30 The two-way analysis of variance involves the same assumptions as other forms of analysis of variance. What are they?

Answer: The populations involved are approximately normally distributed and have approximately equal variances.

31 You expect one of your experimental treatments to have a dramatic effect on the variability of the data. Should you plan a two-way analysis of variance for your statistical test?

Answer: No

32 You are studying measures that in your general experience are bimodally distributed. Should you plan to use analysis of variance for a statistical test?

Answer: No. In this case a χ^2 test would probably be appropriate, since the data seem to fall into two groups.

33 A chemist is studying the production of an organic compound in a process that involves a catalyst. She believes that the amount of the compound produced in a given processing time will vary according to the catalyst used and according to the temperature, but that the relative effectiveness of the different catalysts is not influenced by the temperature. Assume that five catalysts and four temperature levels are to be investigated. The measure to be used will be units of the compound produced in a five-minute processing run. The chemist will make a total of 200 processing runs.

(a) Outline the design for an analysis of variance study. What will the rows and columns be, and how many measurements will be made for each cell?

(b) According to the chemist's theory, which *F* ratios should be significant? Which should not be significant?

Be sure to consider all *F* ratios.

Answer:

(a)

	Heat				Catalyst
	A	B	C	D	E
1					
2					
3	10 measurements per cell				
4					

(b) Both the row and column variances should be significant, but there should be no significant interaction because this would indicate a situation in which the relative effectiveness of the catalysts is influenced by temperature.

34 A researcher believes that he has developed a test of language-learning ability that will predict a student's ability to learn any language. He proposed to evaluate his test by using it to classify students as high, medium, or low in learning ability, giving them language training and then achievement tests. To make sure that the test works uniformly well for several languages, he will train and test students in Spanish, Russian, Swahili, and Japanese. Thus he plans a two-way analysis of variance. The original plan was to train and test each student on only one language, but the researcher ran short of high-ability students. He then trained and tested 10 high-ability students in two languages each. Is this an acceptable procedure?

Answer: No. Analysis of variance assumes independent random samples.

35 Assuming that the researcher obtains an appropriate sample and sets up his study so that the different languages are the columns, which *F* ratios should be significant?

Answer: The *F* ratio for rows (ability level) would be significant. The interaction should *not* be significant, since he expects the test to work equally well for all languages.

36 What would a significant *F* ratio for columns (languages) mean?

Answer: That the achievement tests are not of equal difficulty or that some languages are more difficult to learn than others.

SELF-TEST If you have successfully completed this chapter, you can now perform and interpret a two-way analysis of variance. You can:

• Compute *F* ratios for rows, columns, and interaction

• Examine cell means to interpret the results of two-way analysis of variance

• Recognize situations in which a two-way analysis of variance is not appropriate and suggest alternative tests

Now try these review problems. Table I at the back of the book lists any formulas you may need for reference.

1. A study is conducted of the effects of motivation and method of instruction on achievement in classroom learning. One hundred high school students are assigned at random to four groups. Half the students are promised 50 cents per point on the final exam. The others are paid $5 to participate in the experiment, regardless of how they score. Half the "high motivation" group (the groups paid 50 cents per point) and half the "low motivation" group (the group paid regardless of score) are taught by one instructor, who employs an inductive discovery approach. The remaining students are taught by a second instructor, who employs an expository deductive approach. A standardized final exam is administered to all students. The results of the study are as follows.

	Inductive	Deductive	Total
High motivation	$\bar{x} = 54$	$\bar{x} = 55$	54.5
	($s = 5.1$)	($s = 5.3$)	
Low motivation	$\bar{x} = 56$	$\bar{x} = 35$	45.5
	($s = 4.9$)	($s = 6.1$)	
Total	55	45	

An analysis of variance shows no significant effect of motivation or of method of instruction, but the interaction is significant at the 5% level. The researcher concludes on the basis of these results that the inductive method is superior for poorly motivated students, whereas the deductive method is superior for well-motivated students. Do the data support these conclusions? Why or why not?

2. The following are dexterity scores for boys and girls at three age levels. A high score indicates many errors and low dexterity. Complete a two-way analysis of variance and interpret the results. Refer to the formulas in table I at the back of the book for guidance.

Age	5–7	7–9	9–11
	5	3	1
	6	4	2
Girls	6	5	2
	6	6	2
	7	7	3
	6	3	1
	7	4	2
Boys	7	5	3
	8	6	4
	8	7	5

3. You conclude from a two-way analysis of variance that age appears to be related to dexterity, but gender is not. What statistical techniques could you use to further describe and analyze this relationship?

4. Summarize briefly the assumptions required for:

   (a) A z test for the difference between the means of two samples

   (b) A χ^2 test

   (c) A t test for the difference between the means of two samples

   (d) Analysis of variance

   (e) A test of the null hypothesis $\rho = 0$

ANSWERS

To review a problem, study the frames indicated after the answer.

1. There are two major difficulties with the researcher's interpretation of his results:

(a) The cell means do not fit his description of the interaction. In fact, the interaction appears to consist only of the fact that deductive students with low motivation do less well than all the other groups.

(b) Because each method was used by a single instructor, it is impossible to separate the method of instruction from his other characteristics, such as the loudness of his voice and the radiance of his smile. You can reject the null hypothesis that there is no difference between the groups, but there are numerous plausible alternatives. *See frames 25, 33 to 36, and frames 26 to 31 in chapter 4.*

2.

	5–7		7–9		9–11		Totals	
	x	x^2	x	x^2	x	x^2	x	x^2
	5	25	3	9	1	1		
Girls	6	36	4	16	2	4		
	6	36	5	25	2	4		
	6	36	6	36	2	4		
	7	49	7	49	3	9		
	30	182	25	135	10	22	65	339
	6	36	3	9	1	1		
	7	49	4	16	2	4		
Boys	7	49	5	25	3	9		
	8	64	6	36	4	16		
	8	64	7	49	5	25		
	36	262	25	135.	15	55	76	452
Totals	66	444	50	270.	25	77	141.	791

Total sum of squares:

$$\sum x_T^2 - \frac{(\sum x_T)^2}{N} = 791 - \frac{(141)^2}{30} = 791 - 662.7 = 128.3$$

$$df = N - 1 = 29$$

Between-groups sum of squares:

$$\frac{(\sum x_1)^2}{n_1} + \frac{(\sum x_2)^2}{n_2} + \frac{(\sum x_3)^2}{n_3} + \frac{(\sum x_4)^2}{n_4} + \frac{(\sum x_5)^2}{n_5} + \frac{(\sum x_6)^2}{n_6} - \frac{(\sum x_T)^2}{N}$$

$$= \frac{(30)^2}{5} + \frac{(25)^2}{5} + \frac{(10)^2}{5} + \frac{(36)^2}{5} + \frac{(25)^2}{5} + \frac{(15)^2}{5} - 662.7$$

$$= 180.00 + 125.00 + 20.00 + 259.20 + 125 + 45 - 662.7 = 91.50$$

$$\text{df} = g - 1 = 1$$

Between-rows sum of squares:

$$\frac{(\sum x_{r1})^2}{nc} + \frac{(\sum x_{r2})^2}{nc} + \frac{(\sum x_T)^2}{N} = \frac{(65)^2}{15} + \frac{(76)^2}{15} - 662.7 = 281.67 + 385.07 - 662.7 = 4.04$$

$$\text{df} = r - 1 = 1$$

Between-columns sum of squares:

$$\frac{(\sum x_{c1})^2}{nr} + \frac{(\sum x_{c2})^2}{nr} + \frac{(\sum x_{c3})^2}{nr} - \frac{(\sum x_T)^2}{N} = \frac{(66)^2}{10} + \frac{(50)^2}{10} + \frac{(25)^2}{10} - 662.7$$

$$= 435.6 + 250.0 + 62.5 - 662.7 = 85.40$$

Interaction sum of squares:

Between groups – (between rows + between columns) = 91.50 – (4.04 + 85.40) = 2.06

Within groups sum of squares:

Total – between groups = 128.30 – 91.50 = 36.80

	Sum of Squares	df	Variance Estimate	F
Total	128.30	29		
Rows (gender)	4.04	1	4.040	2.64
Columns (age)	85.40	2	42.700	27.95
Interaction	2.06	2	1.030	0.67
Within groups	36.80	24	1.533	

The age groupings have a significant effect on scores but gender has not. *See frames 16 to 29.*

3. The techniques of correlation and regression would be appropriate—scattergrams, correlation coefficients, and regression equations. *See chapter 7.*

4. (a) Both samples are larger than thirty. *See chapter 5, frames 27 to 31.*

   (b) All cells have predicted frequency greater than 5. *See chapter 8, frames 26 to 30.*

   (c) Both populations are approximately normally distributed and have approximately equal variances. *See chapter 5, frames 32 to 35.*

   (d) All populations are approximately normally distributed and have approximately equal variances. *See chapter 6, frames 73 to 78.*

   (e) Both scores are normally distributed; sample is drawn at random. *See chapter 7, frames 35 to 45.*

Appendix A   Tables

TABLE I   REFERENCE FORMULAS

Parameters and Statistics

$$\mu = \frac{\Sigma x}{n} \qquad \overline{x} = \frac{\Sigma x}{n}$$

$$\sigma = \sqrt{\frac{\Sigma(x - \mu)^2}{n}} \ \text{or} \ \sqrt{\frac{\Sigma x^2 - (\Sigma x)^2/n}{n}}$$

$$s = \sqrt{\frac{\Sigma(x - \overline{x})^2}{n - 1}} \ \text{or} \ \sqrt{\frac{\Sigma x^2 - (\Sigma x)^2/n}{n - 1}}$$

$$z = \frac{\overline{x} - \mu}{\sigma_{\overline{x}}} \ \text{or} \ \frac{\overline{x} - \mu}{\sigma/\sqrt{n}}$$

$$r = \frac{1}{n - 1}\Sigma\frac{(x - \overline{x})}{s_x} \cdot \frac{(y - \overline{y})}{s_y} \ \text{or} \ \frac{n\Sigma(xy) - (\Sigma x)(\Sigma y)}{\sqrt{[n\Sigma x^2 - (\Sigma x)^2][n\Sigma y^2 - (\Sigma y)^2]}}$$

Central Limit Theorem for Large Samples

$$\mu_{\overline{x}} = \mu$$

$$\sigma_{\overline{x}} = \frac{\sigma}{\sqrt{n}}$$

Regression Analysis

$$y = \overline{y} + b(x - \overline{x})$$

$$b = \frac{n\Sigma(xy) - (\Sigma x)(\Sigma y)}{n\Sigma x^2 - (\Sigma x)^2}$$

Confidence Intervals

$$\overline{x} \pm z_0 \frac{s}{\sqrt{n}}$$

$$\overline{x} \pm t_0 \frac{s}{\sqrt{n}}$$

$$p \pm z_0 \frac{\sqrt{pq}}{\sqrt{n}}$$

Hypothesis Testing

Null Hypothesis $\mu = C$

$$z = \frac{\overline{x} - C}{s/\sqrt{n}}$$

$$t = \frac{\overline{x} - C}{s/\sqrt{n}}$$

TABLE I *(continued)*

Null Hypothesis $\mu_1 = \mu_2$

$$z = \frac{\overline{x}_1 - \overline{x}_2}{\sqrt{s_1^2/n_1 + s_2^2/n_2}}$$

$$t = \frac{\overline{x}_1 - \overline{x}_2}{\sqrt{s^2/n_1 + s^2/n_2}} \qquad s^2 = \frac{(n_1 - 1)s_1^2 + (n_2 - 1)s_2^2}{n_1 + n_2 - 2}$$

Null Hypothesis $\sigma_1 = \sigma_2$

$$F = \frac{s_1^2}{s_2^2}$$

Null Hypothesis sample is a random sample from a population with a given distribution.

$$\chi^2 = \frac{(f - F)^2}{F}$$

$$df = c - 1 \text{ or } (c-1)(r-1)$$

Null Hypothesis $\mu_1 = \mu_2 = \mu_3 = $ etc.

ONE-WAY ANALYSIS OF VARIANCE

$$F = \frac{\text{between groups variance}}{\text{within groups variance}}$$

Total sum of squares

$$\Sigma x_T^2 - \frac{(\Sigma x_T)^2}{N}$$

$$df = N - 1$$

Between groups sum of squares

$$\frac{(\Sigma x_1)^2}{n_1} + \frac{(\Sigma x_2)^2}{n_2} + \ldots \text{ etc. } - \frac{(\Sigma x_T)^2}{N}$$

$$df = g - 1$$

Within groups sum of squares
Total sum of squares − between groups sum of squares

or

$$\left[\Sigma x_1^2 - \frac{(\Sigma x_1)^2}{n_1} \right] + \left[\Sigma x_2^2 - \frac{(\Sigma x_2)^2}{n_2} \right] + \ldots \text{ etc.}$$

$$df = N - g$$

TABLE I *(continued)*

TWO-WAY ANALYSIS OF VARIANCE

$$F = \frac{\text{between rows variance}}{\text{within groups variance}}$$

$$F = \frac{\text{between columns variance}}{\text{within groups variance}}$$

$$F = \frac{\text{interaction variance}}{\text{within groups variance}}$$

Total sum of squares

$$\Sigma x_T^2 - \frac{(\Sigma x_T)^2}{N}$$

Between groups sum of squares

$$\frac{(\Sigma x_1)^2}{n} + \frac{(\Sigma x_2)^2}{n} + \dots \text{etc.} - \frac{(\Sigma x_T)^2}{N}$$

$$df = rc - 1$$

Between rows sum of squares

$$\frac{(\Sigma x_{row1})^2}{nc} + \frac{(\Sigma x_{row2})^2}{nc} + \dots \text{etc.} - \frac{(\Sigma x_T)^2}{N}$$

$$df = r - 1$$

Between columns sum of squares

$$\frac{(\Sigma x_{col1})^2}{nr} + \frac{(\Sigma x_{col2})^2}{nr} + \dots \text{etc.} - \frac{(\Sigma x_T)^2}{N}$$

$$df = c - 1$$

Interaction sum of squares

Between groups sum of squares − [*between rows sum of squares* + *between columns sum of squares*]

or

$$\frac{(\Sigma x_1)^2}{n} + \frac{(\Sigma x_2)^2}{n} + \dots \text{etc.} - \frac{(\Sigma x_{row1})^2}{nc} - \frac{(\Sigma x_{row2})^2}{nc} - \dots \text{etc.} - \frac{(\Sigma x_{col1})^2}{nr}$$

$$- \frac{(\Sigma x_{col2})^2}{nr} - \dots \text{etc.} + \frac{(\Sigma x_T)^2}{N}$$

$$df = (r - 1)(c - 1)$$

Within groups sum of squares

Total sum of squares − between groups sum of squares

or

$$\left[\Sigma x_1^2 - \frac{(\Sigma x_1)^2}{n} \right] + \left[\Sigma x_2^2 - \frac{(\Sigma x_2)^2}{n} \right] + \dots \text{etc.}$$

$$df = N - rc$$

HOW TO USE TABLE II

To find the square root of a number between 1 and 10, find your number in the column headed **N**. Look across to the column headed $\sqrt{N}$ for the square root. For example, to find the square root of 5:

    locate 5.00 under **N**

    look across to find $\sqrt{N}$ = 2.23607

To find the square root of a number from 10 to 100, divide the number by 10. Find the result in the column headed **N**. Then look across to the column headed $\sqrt{10N}$ for your square root. For example, to find the square root of 50:

    $50 \div 10 = 5$

    locate 5.00 in the **N** column

    look across to find $\sqrt{10N}$ = 7.07107

To find the square root of a number over 100, move the decimal point an even number of places to the left until you have a number between 1 and 100. Look up the square root of that number in the table, as shown above. Then move the decimal point in the answer back to the right *half* as many places as you moved to the left. For example, to find the square root of 5000:

    move the decimal point *two* places to the left 50.00

    $\sqrt{50}$ = 7.07107

    move the decimal point in the answer *one* place to the right

    70.7107

    $\sqrt{5000}$ = 70.7107

To find the square root of a number less than one, move the decimal point an even number of places to the right until you have a number between 1 and 100. Look up the square root of that number in the table, as shown above. Then move the decimal point in the answer back to the left *half* as many places as you moved to the right. For example, to find the square root of 0.0710:

    move the decimal point *two* places to the right 07.10

    $\sqrt{7.10}$ = 2.66458

    move the decimal point *one* place to the left .266458

    $\sqrt{0.0710}$ = 0.266458

TABLE II   SQUARES AND SQUARE ROOTS

N	N²	√N	√10N	N	N²	√N	√10N
1.00	1.0000	1.00000	3.16228	**1.50**	2.2500	1.22474	3.87298
1.01	1.0201	1.00499	3.17805	1.51	2.2801	1.22882	3.88587
1.02	1.0404	1.00995	3.19374	1.52	2.3104	1.23288	3.89872
1.03	1.0609	1.01489	3.20936	1.53	2.3409	1.23693	3.91152
1.04	1.0816	1.01980	3.22490	1.54	2.3716	1.24097	3.92428
1.05	1.1025	1.02470	3.24037	1.55	2.4025	1.24499	3.93700
1.06	1.1236	1.02956	3.25576	1.56	2.4336	1.24900	3.94968
1.07	1.1449	1.03441	3.27109	1.57	2.4649	1.25300	3.96232
1.08	1.1664	1.03923	3.28634	1.58	2.4964	1.25698	3.97492
1.09	1.1881	1.04403	3.30151	1.59	2.5281	1.26095	3.98748
1.10	1.2100	1.04881	3.31662	**1.60**	2.5600	1.26491	4.00000
1.11	1.2321	1.05357	3.33167	1.61	2.5921	1.26886	4.01248
1.12	1.2544	1.05830	3.34664	1.62	2.6244	1.27279	4.02492
1.13	1.2769	1.06301	3.36155	1.63	2.6569	1.27671	4.03733
1.14	1.2996	1.06771	3.37639	1.64	2.6896	1.28062	4.04969
1.15	1.3225	1.07238	3.39116	1.65	2.7225	1.28452	4.06202
1.16	1.3456	1.07703	3.40588	1.66	2.7556	1.28841	4.07431
1.17	1.3689	1.08167	3.42053	1.67	2.7889	1.29228	4.08656
1.18	1.3924	1.08628	3.43511	1.68	2.8224	1.29615	4.09878
1.19	1.4161	1.09087	3.44964	1.69	2.8561	1.30000	4.11096
1.20	1.4400	1.09545	3.46410	**1.70**	2.8900	1.30384	4.12311
1.21	1.4641	1.10000	3.47851	1.71	2.9241	1.30767	4.13521
1.22	1.4884	1.10454	3.49285	1.72	2.9584	1.31149	4.14729
1.23	1.5129	1.10905	3.50714	1.73	2.9929	1.31529	4.15933
1.24	1.5376	1.11355	3.52136	1.74	3.0276	1.31909	4.17133
1.25	1.5625	1.11803	3.53553	1.75	3.0625	1.32288	4.18330
1.26	1.5876	1.12250	3.54965	1.76	3.0976	1.32665	4.19524
1.27	1.6129	1.12694	3.56371	1.77	3.1329	1.33041	4.20714
1.28	1.6384	1.13137	3.57771	1.78	3.1684	1.33417	4.21900
1.29	1.6641	1.13578	3.59166	1.79	3.2041	1.33791	4.23084
1.30	1.6900	1.14018	3.60555	**1.80**	3.2400	1.34164	4.24264
1.31	1.7161	1.14455	3.61939	1.81	3.2761	1.34536	4.25441
1.32	1.7424	1.14891	3.63318	1.82	3.3124	1.34907	4.26615
1.33	1.7689	1.15326	3.64692	1.83	3.3489	1.35277	4.27785
1.34	1.7956	1.15758	3.66060	1.84	3.3856	1.35647	4.28952
1.35	1.8225	1.16190	3.67423	1.85	3.4225	1.36015	4.30116
1.36	1.8496	1.16619	3.68782	1.86	3.4596	1.36382	4.31277
1.37	1.8769	1.17047	3.70135	1.87	3.4969	1.36748	4.32435
1.38	1.9044	1.17473	3.71484	1.88	3.5344	1.37113	4.33590
1.39	1.9321	1.17898	3.72827	1.89	3.5721	1.37477	4.34741
1.40	1.9600	1.18322	3.74166	**1.90**	3.6100	1.37840	4.35890
1.41	1.9881	1.18743	3.75500	1.91	3.6481	1.38203	4.37035
1.42	2.0164	1.19164	3.76829	1.92	3.6864	1.38564	4.38178
1.43	2.0449	1.19583	3.78153	1.93	3.7249	1.38924	4.39318
1.44	2.0736	1.20000	3.79473	1.94	3.7636	1.39284	4.40454
1.45	2.1025	1.20416	3.80789	1.95	3.8025	1.39642	4.41588
1.46	2.1316	1.20830	3.82099	1.96	3.8416	1.40000	4.42719
1.47	2.1609	1.21244	3.83406	1.97	3.8809	1.40357	4.43847
1.48	2.1904	1.21655	3.84708	1.98	3.9204	1.40712	4.44972
1.49	2.2201	1.22066	3.86005	1.99	3.9601	1.41067	4.46094
1.50	2.2500	1.22474	3.87298	**2.00**	4.0000	1.41421	4.47214
N	N²	√N	√10N	N	N²	√N	√10N

TABLE II *(continued)*

N	N²	√N̄	√10N̄	N	N²	√N̄	√10N̄
2.00	4.0000	1.41421	4.47214	**2.50**	6.2500	1.58114	5.00000
2.01	4.0401	1.41774	4.48330	2.51	6.3001	1.58430	5.00999
2.02	4.0804	1.42127	4.49444	2.52	6.3504	1.58745	5.01996
2.03	4.1209	1.42478	4.50555	2.53	6.4009	1.59060	5.02991
2.04	4.1616	1.42829	4.51664	2.54	6.4516	1.59374	5.03984
2.05	4.2025	1.43178	4.52769	2.55	6.5025	1.59687	5.04975
2.06	4.2436	1.43527	4.53872	2.56	6.5536	1.60000	5.05964
2.07	4.2849	1.43875	4.54973	2.57	6.6049	1.60312	5.06952
2.08	4.3264	1.44222	4.56070	2.58	6.6564	1.60624	5.07937
2.09	4.3681	1.44568	4.57165	2.59	6.7081	1.60935	5.08920
2.10	4.4100	1.44914	4.58258	**2.60**	6.7600	1.61245	5.09902
2.11	4.4521	1.45258	4.59347	2.61	6.8121	1.61555	5.10882
2.12	4.4944	1.45602	4.60435	2.62	6.8644	1.61864	5.11859
2.13	4.5369	1.45945	4.61519	2.63	6.9169	1.62173	5.12835
2.14	4.5796	1.46287	4.62601	2.64	6.9696	1.62481	5.13809
2.15	4.6225	1.46629	4.63681	2.65	7.0225	1.62788	5.14782
2.16	4.6656	1.46969	4.64758	2.66	7.0756	1.63095	5.15752
2.17	4.7089	1.47309	4.65833	2.67	7.1289	1.63401	5.16720
2.18	4.7524	1.47648	4.66905	2.68	7.1824	1.63707	5.17687
2.19	4.7961	1.47986	4.67974	2.69	7.2361	1.64012	5.18652
2.20	4.8400	1.48324	4.69042	**2.70**	7.2900	1.64317	5.19615
2.21	4.8841	1.48661	4.70106	2.71	7.3441	1.64621	5.20577
2.22	5.9284	1.48997	4.71169	2.72	7.3984	1.64924	5.21536
2.23	4.9729	1.49332	4.72229	2.73	7.4529	1.65227	5.22494
2.24	5.0176	1.49666	4.73286	2.74	7.5076	1.65529	5.23450
2.25	5.0625	1.50000	4.74342	2.75	7.5625	1.65831	5.24404
2.26	5.1076	1.50333	4.75395	2.76	7.6176	1.66132	5.25357
2.27	5.1529	1.50665	4.76445	2.77	7.6729	1.66433	5.26308
2.28	5.1984	1.50997	4.77493	2.78	7.7284	1.66733	5.27257
2.29	5.2441	1.51327	4.78539	2.79	7.7841	1.67033	5.28205
2.30	5.2900	1.51658	4.79583	**2.80**	7.8400	1.67332	5.29150
2.31	5.3361	1.51987	4.80625	2.81	7.8961	1.67631	5.30094
2.32	5.3824	1.52315	4.81664	2.82	7.9524	1.67929	5.31037
2.33	5.4289	1.52643	4.82701	2.83	8.0089	1.68226	5.31977
2.34	5.4756	1.52971	4.83735	2.84	8.0656	1.68523	5.32917
2.35	5.5225	1.53297	4.84768	2.85	8.1225	1.68819	5.33854
2.36	5.5696	1.53623	4.85798	2.86	8.1796	1.69115	5.34790
2.37	5.6169	1.53948	4.86826	2.87	8.2369	1.69411	5.35724
2.38	5.6644	1.54272	4.87852	2.88	8.2944	1.69706	5.36656
2.39	5.7121	1.54596	4.88876	2.89	8.3521	1.70000	5.37587
2.40	5.7600	1.54919	4.89898	**2.90**	8.4100	1.70294	5.38516
2.41	5.8081	1.55252	4.90918	2.91	8.4681	1.70587	5.39444
2.42	5.8564	1.55563	4.91935	2.92	8.5264	1.70880	5.40370
2.43	5.9049	1.55885	4.92950	2.93	8.5849	1.71172	5.41295
2.44	5.9536	1.56205	4.93964	2.94	8.6436	1.71464	5.42218
2.45	6.0025	1.56525	4.94975	2.95	8.7025	1.71756	5.43139
2.46	6.0516	1.56844	4.95984	2.96	8.7616	1.72047	5.44059
2.47	6.1009	1.57162	4.96991	2.97	8.8209	1.72337	5.44977
2.48	6.1054	1.57480	4.97996	2.98	8.8804	1.72627	5.45894
2.49	6.2001	1.57797	4.98999	2.99	8.9401	1.72916	5.46809
2.50	6.2500	1.58114	5.00000	**3.00**	9.0000	1.73205	5.47723
N	N²	√N̄	√10N̄	N	N²	√N̄	√10N̄

TABLE II   *(continued)*

N	N²	$\sqrt{N}$	$\sqrt{10N}$	N	N²	$\sqrt{N}$	$\sqrt{10N}$
3.00	9.0000	1.73205	5.47723	**3.50**	12.2500	1.87083	5.91608
3.01	9.0601	1.73494	5.48635	3.51	12.3201	1.87350	5.92453
3.02	9.1204	1.73781	5.49545	3.52	12.3904	1.87617	5.93296
3.03	9.1809	1.74069	5.50454	3.53	12.4609	1.87883	5.94138
3.04	9.2416	1.74356	5.51362	3.54	12.5316	1.88149	5.94979
3.05	9.3025	1.74642	5.52268	3.55	12.6025	1.88414	5.95819
3.06	9.3636	1.74929	5.53173	3.56	12.6736	1.88680	5.96657
3.07	9.4249	1.75214	5.54076	3.57	12.7449	1.88944	5.97495
3.08	9.4864	1.75499	5.54977	3.58	12.8164	1.89209	5.98331
3.09	9.5481	1.75784	5.55878	3.59	12.8881	1.89473	5.99166
3.10	9.6100	1.76068	5.56776	**3.60**	12.9600	1.89737	6.00000
3.11	9.6721	1.76352	5.57674	3.61	13.0321	1.90000	6.00833
3.12	9.7344	1.76635	5.58570	3.62	13.1044	1.90263	6.01664
3.13	9.7969	1.76918	5.59464	3.63	13.1769	1.90526	6.02495
3.14	9.8596	1.77200	5.60357	3.64	13.2496	1.90788	6.03324
3.15	9.9225	1.77482	5.61249	3.65	13.3225	1.91050	6.04152
3.16	9.9856	1.77764	5.62139	3.66	13.3956	1.91311	6.04949
3.17	10.0489	1.78045	5.63028	3.67	13.4689	1.91572	6.05805
3.18	10.1124	1.78326	5.63915	3.68	13.5424	1.91833	6.06630
3.19	10.1761	1.78606	5.64801	3.69	13.6161	1.92094	6.07454
3.20	10.2400	1.78885	5.65685	**3.70**	13.6900	1.92354	6.08276
3.21	10.3041	1.79165	5.66569	3.71	13.7641	1.92614	6.09098
3.22	10.3684	1.79444	5.67450	3.72	13.8384	1.92873	6.09918
3.23	10.4329	1.79722	5.68331	3.73	13.9129	1.93132	6.10737
3.24	10.4976	1.80000	5.69210	3.74	13.9876	1.93391	6.11555
3.25	10.5625	1.80278	5.70088	3.75	14.0625	1.93649	6.12372
3.26	10.6276	1.80555	5.70964	3.76	14.1376	1.93907	6.13188
3.27	10.6929	1.80831	5.71839	3.77	14.2129	1.94165	6.14003
3.28	10.7584	1.81108	5.72713	3.78	14.2884	1.94422	6.14817
3.29	10.8241	1.81384	5.73585	3.79	14.3641	1.94679	6.15630
3.30	10.8900	1.81659	5.74456	**3.80**	14.4400	1.94936	6.16441
3.31	10.9561	1.81934	5.75326	3.81	14.5161	1.95192	6.17252
3.32	10.0224	1.82209	5.76194	3.82	14.5924	1.95448	6.18061
3.33	11.0889	1.82483	5.77062	3.83	14.6689	1.95704	6.18870
3.34	11.1556	1.82757	5.77927	3.84	14.7456	1.95959	6.19677
3.35	11.2225	1.83030	5.78792	3.85	14.8225	1.96214	6.20484
3.36	11.2896	1.83303	5.79655	3.86	14.8996	1.96469	6.21289
3.37	11.3569	1.83576	5.80517	3.87	14.9769	1.96723	6.22093
3.38	11.4244	1.83848	5.81378	3.88	15.0544	1.96977	6.22896
3.39	11.4921	1.84120	5.82237	3.89	15.1321	1.97231	6.23699
3.40	11.5600	1.84391	5.83095	**3.90**	15.2100	1.97484	6.24500
3.41	11.6281	1.84662	5.83952	3.91	15.2881	1.97737	6.25300
3.42	11.6964	1.84932	5.84808	3.92	15.3664	1.97990	6.26099
3.43	11.7649	1.85203	5.85662	3.93	15.4449	1.98242	6.26897
3.44	11.8336	1.85472	5.86515	3.94	15.5236	1.98494	6.27694
3.45	11.9025	1.85742	5.87367	3.95	15.6025	1.98746	6.28490
3.46	11.9716	1.86011	5.88218	3.96	15.6816	1.98997	6.29285
3.47	12.0409	1.86279	5.89067	3.97	15.7609	1.99249	6.30079
3.48	12.1104	1.86548	5.89915	3.98	15.8404	1.99499	6.30872
3.49	12.1801	1.86815	5.90762	3.99	15.9201	1.99750	6.31644
3.50	12.2500	1.87083	5.91608	**4.00**	16.0000	2.00000	6.32456
N	N²	$\sqrt{N}$	$\sqrt{10N}$	N	N²	$\sqrt{N}$	$\sqrt{10N}$

TABLE II *(continued)*

N	N²	√N	√10N	N	N²	√N	√10N
4.00	16.0000	2.00000	6.32456	**4.50**	20.2500	2.12132	6.70820
4.01	16.0801	2.00250	6.33246	4.51	20.3401	2.12368	6.71565
4.02	16.1604	2.00499	6.34035	4.52	20.4304	2.12603	6.72309
4.03	16.2409	2.00749	6.34823	4.53	20.5209	2.12838	6.73053
4.04	16.3216	2.00998	6.35610	4.54	20.6116	2.13073	6.73795
4.05	16.4025	2.01246	6.36396	4.55	20.7025	2.13307	6.74537
4.06	16.4836	2.01494	6.37181	4.56	20.7936	2.13542	6.75278
4.07	16.5649	2.01742	6.37966	4.57	20.8849	2.13776	6.76018
4.08	16.6464	2.01990	6.38749	4.58	20.9764	2.14009	6.76757
4.09	16.7281	2.02237	6.39531	4.59	21.0681	2.14243	6.77495
4.10	16.8100	2.02485	6.40312	**4.60**	21.1600	2.14476	6.78233
4.11	16.8921	2.02731	6.41093	4.61	21.2521	2.14709	6.78970
4.12	16.9744	2.02978	6.41872	4.62	21.3444	2.14942	6.79706
4.13	17.0569	2.03224	6.42651	4.63	21.4369	2.15174	6.80441
4.14	17.1396	2.03470	6.43428	4.64	21.5296	2.15407	6.81175
4.15	17.2225	2.03715	6.44205	4.65	21.6225	2.15639	6.81909
4.16	17.3056	2.03961	6.44981	4.66	21.7156	2.15870	6.82642
4.17	17.3889	2.04206	6.45755	4.67	21.8089	2.16102	6.83374
4.18	17.4724	2.04450	6.46529	4.68	21.9024	2.16333	6.84105
4.19	17.5561	2.04695	6.47302	4.69	21.9961	2.16564	6.84836
4.20	17.6400	2.04939	6.48074	**4.70**	22.0900	2.16795	6.85565
4.21	17.7241	2.05183	6.48845	4.71	22.1841	2.17025	6.86294
4.22	17.8084	2.05426	6.49615	4.72	22.2784	2.17256	6.87023
4.23	17.8929	2.05670	6.50384	4.73	22.3729	2.17486	6.87750
4.24	17.9776	2.05913	6.51153	4.74	22.4676	2.17715	6.88477
4.25	18.0625	2.06155	6.51920	4.75	22.5625	2.17945	6.89202
4.26	18.1476	2.06398	6.52687	4.76	22.6576	2.18174	6.89928
4.27	18.2329	2.06640	6.53452	4.77	22.7529	2.18403	6.90652
4.28	18.3184	2.06882	6.54217	4.78	22.8484	2.18632	6.91375
4.29	18.4041	2.07123	6.54981	4.79	22.9441	2.18861	6.92098
4.30	18.4900	2.07364	6.55744	**4.80**	23.0400	2.19089	6.92820
4.31	18.5761	2.07605	6.66506	4.81	23.1361	2.19317	6.93542
4.32	18.6624	2.07846	6.57267	4.82	23.2324	2.19545	6.94262
4.33	18.7489	2.08087	6.58027	4.83	23.3289	2.19773	6.94982
4.34	18.8356	2.08327	6.58787	4.84	23.4256	2.20000	6.95701
4.35	18.9225	2.08567	6.59545	4.85	23.5225	2.20227	6.96419
4.36	19.0096	2.08806	6.60303	4.86	23.6196	2.20454	6.97137
4.37	19.0969	2.09045	6.61060	4.87	23.7169	2.20681	6.97854
4.38	19.1844	2.09284	6.61816	4.88	23.8144	2.20907	6.98570
4.39	19.2721	2.09523	6.62571	4.89	23.9121	2.21133	6.99285
4.40	19.3600	2.09762	6.63325	**4.90**	24.0100	2.21359	7.00000
4.41	19.4481	2.10000	6.64078	4.91	24.1081	2.21585	7.00714
4.42	19.5364	2.10238	6.64831	4.92	24.2064	2.21811	7.01427
4.43	19.6249	2.10476	6.65582	4.93	24.3049	2.22036	7.02140
4.44	19.7136	2.10713	6.66333	4.94	24.4036	2.22261	7.02851
4.45	19.8025	2.10950	6.67083	4.95	24.5025	2.22486	7.03562
4.46	19.8916	2.11187	6.67832	4.96	24.6016	2.22711	7.04273
4.47	19.9809	2.11424	6.68581	4.97	24.7009	2.22935	7.04982
4.48	20.0704	2.11660	6.69328	4.98	24.8004	2.23159	7.05691
4.49	20.1601	2.11896	6.70075	4.99	24.9001	2.23383	7.06399
4.50	20.2500	2.12132	6.70820	**5.00**	25.0000	2.23607	7.07107
N	N²	√N	√10N	N	N²	√N	√10N

TABLE II *(continued)*

N	N²	√N	√10N	N	N²	√N	√10N
5.00	25.0000	2.23607	7.07107	**5.50**	30.2500	2.34521	7.41620
5.01	25.1001	2.23830	7.07814	5.51	30.3601	2.34734	7.42294
5.02	25.2004	2.24054	7.08520	5.52	30.4704	2.34947	7.42967
5.03	25.3009	2.24277	7.09225	5.53	30.5809	2.35160	7.43640
5.04	25.4016	2.24499	7.09930	5.54	30.6916	2.35372	7.44312
5.05	25.5025	2.24722	7.10634	5.55	30.8025	2.35584	7.44983
5.06	25.6036	2.24944	7.11337	5.56	30.9136	2.35797	7.45654
5.07	25.7049	2.25167	7.12039	5.57	31.0249	2.36008	7.46324
5.08	25.8064	2.25389	7.12741	5.58	31.1364	2.36220	7.46994
5.09	25.9081	2.25610	7.13442	5.59	31.2481	2.36432	7.47663
5.10	26.0100	2.25832	7.14143	**5.60**	31.3600	2.36643	7.48331
5.11	26.1121	2.26053	7.14843	5.61	31.4721	2.36854	7.48999
5.12	26.2144	2.26274	7.15542	5.62	31.5844	2.37065	7.49667
5.13	26.3169	2.26495	7.16240	5.63	31.6969	2.37276	7.50333
5.14	26.4196	2.26716	7.16938	5.64	31.8096	2.37487	7.50999
5.15	26.5225	2.26936	7.17635	5.65	31.9225	2.37697	7.51665
5.16	26.6256	2.27156	7.18331	5.66	32.0356	2.37908	7.52330
5.17	26.7289	2.27376	7.19027	5.67	32.1489	2.38118	7.52994
5.18	26.8324	2.27596	7.19722	5.68	32.2624	2.38328	7.53658
5.19	26.9361	2.27816	7.20417	5.69	32.3761	2.38537	7.54321
5.20	27.0400	2.28035	7.21110	**5.70**	32.4900	2.38747	7.54983
5.21	27.1441	2.28254	7.21803	5.71	32.6041	2.38956	7.55645
5.22	27.2484	2.28473	7.22496	5.72	32.7184	2.39165	7.56307
5.23	27.3529	2.28692	7.23187	5.73	32.8329	2.39374	7.56968
5.24	27.4576	2.28910	7.23838	5.74	32.9476	2.39583	7.57628
5.25	27.5625	2.29129	7.24569	5.75	33.0625	2.39792	7.58288
5.26	27.6676	2.29347	7.25259	5.76	33.1776	2.40000	7.58947
5.27	27.7729	2.29565	7.25948	5.77	33.2929	2.40208	7.59605
5.28	27.8784	2.29783	7.26636	5.78	33.4084	2.40416	7.60263
5.29	27.9841	2.30000	7.27324	5.79	33.5241	2.40624	7.60920
5.30	28.0900	2.30217	7.28011	**5.80**	33.6400	2.40832	7.61577
5.31	28.1961	2.30434	7.28697	5.81	33.7561	2.41039	7.62234
5.32	28.3024	2.30651	7.29383	5.82	33.8724	2.41247	7.62889
5.33	28.4089	2.30868	7.30068	5.83	33.9889	2.41454	7.63544
5.34	28.5156	2.31084	7.30753	5.84	34.1056	2.41661	7.64199
5.35	28.6225	2.31301	7.31437	5.85	34.2225	2.41868	7.64853
5.36	28.7296	2.31517	7.32120	5.86	34.3396	2.42074	7.65506
5.37	28.8369	2.31733	7.32803	5.87	34.4569	2.42281	7.66159
5.38	28.9444	2.31948	7.33485	5.88	34.5744	2.42487	7.66812
5.39	29.0521	2.32164	7.34166	5.89	34.6921	2.42693	7.67463
5.40	29.1600	2.32379	7.34847	**5.90**	34.8100	2.42899	7.68115
5.41	29.2681	2.32594	7.35527	5.91	34.9281	2.43105	7.68765
5.42	29.3764	2.32809	7.36206	5.92	35.0464	2.43311	7.69415
5.43	29.4849	2.33024	7.36885	5.93	35.1649	2.43516	7.70065
5.44	29.5936	2.33238	7.37564	5.94	35.2836	2.43721	7.70714
5.45	29.7025	2.33452	7.38241	5.95	35.4025	2.43926	7.71362
5.46	29.8116	2.33666	7.38918	5.96	35.5216	2.44131	7.72010
5.47	29.9209	2.33880	7.39594	5.97	35.6409	2.44336	7.72658
5.48	30.0304	2.34094	7.40270	5.98	35.7604	2.44540	7.73305
5.49	30.1401	2.34307	7.40945	5.99	35.8801	2.44745	7.73951
5.50	30.2500	2.34521	7.41620	**6.00**	36.0000	2.44949	7.74597
N	N²	√N	√10N	N	N²	√N	√10N

TABLE II *(continued)*

N	N²	√N	√10N	N	N²	√N	√10N
6.00	36.0000	2.44949	7.74597	**6.50**	42.2500	2.54951	8.06226
6.01	36.1201	2.45153	7.75242	6.51	42.3801	2.55147	8.06846
6.02	36.2404	2.45357	7.75887	6.52	42.5104	2.55343	8.07465
6.03	36.3609	2.45561	7.76531	6.53	42.6409	2.55539	8.08084
6.04	36.4816	2.45764	7.77174	6.54	42.7716	2.55734	8.08703
6.05	36.6025	2.45967	7.77817	6.55	42.9025	2.55930	8.09321
6.06	36.7236	2.46171	7.78460	6.56	43.0336	2.56125	8.09938
6.07	36.8449	2.46374	7.79102	6.57	43.1649	2.56320	8.10555
6.08	36.9664	2.46577	7.79744	6.58	43.2964	2.56515	8.11172
6.09	37.0881	2.46779	7.80385	6.59	43.4281	2.56710	8.11788
6.10	37.2100	2.46982	7.81025	**6.60**	43.5600	2.56905	8.12404
6.11	37.3321	2.47184	7.81665	6.61	43.6921	2.57099	8.13019
6.12	37.4544	2.47386	7.82304	6.62	43.8244	2.57294	8.13634
6.13	37.5769	2.47588	7.82943	6.63	43.9569	2.57488	8.14248
6.14	37.6996	2.47790	7.83582	6.64	44.0896	2.57682	8.14862
6.15	37.8225	2.47992	7.84219	6.65	44.2225	2.57876	8.15475
6.16	37.9456	2.48193	7.84857	6.66	44.3556	2.58070	8.16088
6.17	38.0689	2.48395	7.85493	6.67	44.4889	2.58263	8.16701
6.18	38.1924	2.48596	7.86130	6.68	44.6224	2.58457	8.17313
6.19	38.3161	2.48797	7.86766	6.69	44.7561	2.58650	8.17924
6.20	38.4400	2.48998	7.87401	**6.70**	44.8900	2.58844	8.18535
6.21	38.5641	2.49199	7.88036	6.71	45.0241	2.59037	8.19146
6.22	38.6884	2.49399	7.88670	6.72	45.1584	2.59230	8.19756
6.23	38.8129	2.49600	7.89303	6.73	45.2929	2.59422	8.20366
6.24	38.9376	2.49800	7.89937	6.74	45.4276	2.59615	8.20975
6.25	39.0625	2.50000	7.90569	6.75	45.5625	2.59808	8.21584
6.26	39.1876	2.50200	7.91202	6.76	45.6976	2.60000	8.22192
6.27	39.3129	2.50400	7.91833	6.77	45.8329	2.60192	8.22800
6.28	39.4384	2.50599	7.92465	6.78	45.9684	2.60384	8.23408
6.29	39.5641	2.50799	7.93095	6.79	46.1041	2.60576	8.24015
6.30	39.6900	2.50998	7.93725	**6.80**	46.2400	2.60768	8.24621
6.31	39.8161	2.51197	7.94355	6.81	46.3761	2.60960	8.25227
6.32	39.9424	2.51396	7.94984	6.82	46.5124	2.61151	8.25833
6.33	40.0689	2.51595	7.95613	6.83	46.6489	2.61343	8.26438
6.34	40.1956	2.51794	7.96241	6.84	46.7856	2.61534	8.27043
6.35	40.3225	2.51992	7.96869	6.85	46.9225	2.61725	8.27647
6.36	40.4496	2.52190	7.97496	6.86	47.0596	2.61916	8.28251
6.37	40.5769	2.52389	7.98123	6.87	47.1969	2.62107	8.28855
6.38	40.7044	2.52587	7.98749	6.88	47.3344	2.62298	8.29458
6.39	40.8321	2.52784	7.99375	6.89	47.4721	2.62488	8.30060
6.40	40.9600	2.52982	8.00000	**6.90**	47.6100	2.62679	8.30662
6.41	41.0881	2.53180	8.00625	6.91	47.7481	2.62869	8.31264
6.42	41.2164	2.53377	8.01249	6.92	47.8864	2.63059	8.31865
6.43	41.3449	2.53574	8.01873	6.93	48.0249	2.63249	8.32466
6.44	41.4736	2.53772	8.02496	6.94	48.1636	2.63439	8.33067
6.45	41.6025	2.53969	8.03119	6.95	48.3025	2.63629	8.33667
6.46	41.7316	2.54165	8.03741	6.96	48.4416	2.63818	8.34266
6.47	41.8609	2.54362	8.04363	6.97	48.5809	2.64008	8.34865
6.48	41.9904	2.54558	8.04984	6.98	48.7204	2.64197	8.35464
6.49	42.1201	2.54755	8.05605	6.99	48.8601	2.64386	8.36062
6.50	42.2500	2.54951	8.06226	**7.00**	49.0000	2.64575	8.36660
N	N²	√N	√10N	N	N²	√N	√10N

TABLE II *(continued)*

N	N²	√N	√10N		N	N²	√N	√10N
7.00	49.0000	2.64575	8.36660		**7.50**	56.2500	2.73861	8.66025
7.01	49.1401	2.64764	8.37257		7.51	56.4001	2.74044	8.66603
7.02	49.2804	2.64953	8.37854		7.52	56.5504	2.74226	8.67179
7.03	49.4209	2.65141	8.38451		7.53	56.7009	2.74408	8.67756
7.04	49.5616	2.65330	8.39047		7.54	56.8516	2.74591	8.68332
7.05	49.7025	2.65518	8.39643		7.55	57.0025	2.74773	8.68907
7.06	49.8436	2.65707	8.40238		7.56	57.1536	2.74955	8.69483
7.07	49.9849	2.65895	8.40833		7.57	57.3049	2.75136	8.70057
7.08	50.1264	2.66083	8.41427		7.58	57.4564	2.75318	8.70632
7.09	50.2681	2.66271	8.42021		7.59	57.6081	2.75500	8.71206
7.10	50.4100	2.66458	8.42615		**7.60**	57.7600	2.75681	8.71780
7.11	50.5521	2.66646	8.43208		7.61	57.9121	2.75862	8.72353
7.12	50.6944	2.66833	8.43801		7.62	58.0644	2.76043	8.72926
7.13	50.8369	2.67021	8.44393		7.63	58.2169	2.76225	8.73499
7.14	50.9796	2.67208	8.44985		7.64	58.3696	2.76405	8.74071
7.15	51.1225	2.67395	8.45577		7.65	58.5225	2.76586	8.74643
7.16	51.2656	2.67582	8.46168		7.66	58.6756	2.76767	8.75214
7.17	51.4089	2.67769	8.46759		7.67	58.8289	2.76948	8.75785
7.18	51.5524	2.67955	8.47349		7.68	58.9824	2.77128	8.76356
7.19	51.6961	2.68142	8.47939		7.69	59.1361	2.77308	8.76926
7.20	51.8400	2.68328	8.48528		**7.70**	59.2900	2.77489	8.77496
7.21	51.9841	2.68514	8.49117		7.71	59.4441	2.77669	8.78066
7.22	52.1284	2.68701	8.49706		7.72	59.5984	2.77849	8.78635
7.23	52.2729	2.68887	8.50294		7.73	59.7529	2.78029	8.79204
7.24	52.4176	2.69072	8.50882		7.74	59.9076	2.78209	8.79773
7.25	52.5625	2.69258	8.51469		7.75	60.0625	2.78388	8.80341
7.26	52.7076	2.69444	8.52056		7.76	60.2176	2.78568	8.80909
7.27	52.8529	2.69629	8.52643		7.77	60.3729	2.78747	8.81476
7.28	52.9984	2.69815	8.53229		7.78	60.5284	2.78927	8.82043
7.29	53.1441	2.70000	8.53815		7.79	60.6841	2.79106	8.82610
7.30	53.2900	2.70185	8.54400		**7.80**	60.8400	2.79285	8.83176
7.31	53.4361	2.70370	8.54985		7.81	60.9961	2.79464	8.83742
7.32	53.5824	2.70555	8.55570		7.82	61.1524	2.79643	8.84308
7.33	53.7289	2.70740	8.56154		7.83	61.3089	2.79821	8.84873
7.34	53.8756	2.70924	8.56738		7.84	61.4656	2.80000	8.85438
7.35	54.0225	2.71109	8.57321		7.85	61.6225	2.80179	8.86002
7.36	54.1696	2.71293	8.57904		7.86	61.7796	2.80357	8.86566
7.37	54.3169	2.71477	8.58487		7.87	61.9369	2.80535	8.87130
7.38	54.4644	2.71662	8.59069		7.88	62.0944	2.80713	8.87694
7.39	54.6121	2.71846	8.59651		7.89	62.2521	2.80891	8.88257
7.40	54.7600	2.72029	8.60233		**7.90**	62.4100	2.81069	8.88819
7.41	54.9081	2.72213	8.60814		7.91	62.5681	2.81247	8.89382
7.42	55.0564	2.72397	8.61394		7.92	62.7264	2.81425	8.89944
7.43	55.2049	2.72580	8.61974		7.93	62.8849	2.81603	8.90505
7.44	55.3536	2.72764	8.62554		7.94	63.0436	2.81780	8.91067
7.45	55.5025	2.72947	8.63134		7.95	63.2025	2.81957	8.91628
7.46	55.6516	2.73130	8.63713		7.96	63.3616	2.82135	8.92188
7.47	55.8009	2.73313	8.64292		7.97	63.5209	2.82312	8.92749
7.48	55.9504	2.73496	8.64870		7.98	63.6804	2.82489	8.93308
7.49	56.1001	2.73679	8.65448		7.99	63.8401	2.82666	8.93868
7.50	56.2500	2.73861	8.66025		**8.00**	64.0000	2.82843	8.94427
N	N²	√N	√10N		N	N²	√N	√10N

TABLE II *(continued)*

N	N²	√N	√10N	N	N²	√N	√10N
8.00	64.0000	2.82843	8.94427	**8.50**	72.2500	2.91548	9.21954
8.01	64.1601	2.83019	8.94986	8.51	72.4201	2.91719	9.22497
8.02	64.3204	2.83196	8.95545	8.52	72.5904	2.91890	9.23038
8.03	64.4809	2.83373	8.96103	8.53	72.7609	2.92062	9.23580
8.04	64.6416	2.83549	8.96660	8.54	72.9316	2.92233	9.24121
8.05	64.8025	2.83725	8.97218	8.55	73.1025	2.92404	9.24662
8.06	64.9636	2.83901	8.97775	8.56	73.2736	2.92575	9.25203
8.07	65.1249	2.84077	8.98332	8.57	73.4449	2.92746	9.25743
8.08	65.2864	2.84253	8.98888	8.58	73.6164	2.92916	9.26283
8.09	65.4481	2.84429	8.99444	8.59	73.7881	2.93087	9.26823
8.10	65.6100	2.84605	9.00000	**8.60**	73.9600	2.93258	9.27362
8.11	65.7721	2.84781	9.00555	8.61	74.1321	2.93428	9.27901
8.12	65.9344	2.84956	9.01110	8.62	74.3044	2.93598	9.28440
8.13	66.0969	2.85132	9.01665	8.63	74.4769	2.93769	9.28978
8.14	66.2596	2.85307	9.02219	8.64	74.6496	2.93939	9.29516
8.15	66.4225	2.85482	9.02774	8.65	74.8225	2.94109	9.30054
8.16	66.5856	2.85657	9.03327	8.66	74.9956	2.94279	9.30591
8.17	66.7489	2.85832	9.03881	8.67	75.1689	2.94449	9.31128
8.18	66.9124	2.86007	9.04434	8.68	75.3424	2.94618	9.31665
8.19	67.0761	2.86182	9.04986	8.69	75.5161	2.94788	9.32202
8.20	67.2400	2.86356	9.05539	**8.70**	75.6900	2.94958	9.32738
8.21	67.4041	2.86531	9.06091	8.71	75.8641	2.95127	9.33274
8.22	67.5684	2.86705	9.06642	8.72	76.0384	2.95296	9.33809
8.23	67.7329	2.86880	9.07193	8.73	76.2129	2.95466	9.34345
8.24	67.8976	2.87054	9.07744	8.74	76.3876	2.95635	9.34880
8.25	68.0625	2.87228	9.08295	8.75	76.5625	2.95804	9.35414
8.26	68.2276	2.87402	9.08845	8.76	76.7376	2.95973	9.35949
8.27	68.3929	2.87576	9.09395	8.77	76.9129	2.96142	9.36483
8.28	68.5584	2.87750	9.09945	8.78	77.0884	2.96311	9.37017
8.29	68.7241	2.87924	9.10494	8.79	77.2641	2.96479	9.37550
8.30	68.8900	2.88097	9.11045	**8.80**	77.4400	2.96648	9.38083
8.31	69.0561	2.88271	9.11592	8.81	77.6161	2.96816	9.38616
8.32	69.2224	2.88444	9.12140	8.82	77.7924	2.96985	9.39149
8.33	69.3889	2.88617	9.12688	8.83	77.9689	2.97153	9.39681
8.34	69.5556	2.88791	9.13236	8.84	78.1456	2.97321	9.40213
8.35	69.7225	2.88964	9.13783	8.85	78.3225	2.97489	9.40744
8.36	69.8896	2.89137	9.14330	8.86	78.4996	2.97658	9.41276
8.37	70.0569	2.89310	9.14877	8.87	78.6769	2.97825	9.41807
8.38	70.2244	2.89482	9.15423	8.88	78.8544	2.97993	9.42338
8.39	70.3921	2.89655	9.15969	8.89	79.0321	2.98161	9.42868
8.40	70.5600	2.89828	9.16515	**8.90**	79.2100	2.98329	9.43398
8.41	70.7281	2.90000	9.17061	8.91	79.3881	2.98496	9.43928
8.42	70.8964	2.90172	9.17606	8.92	79.5664	2.98664	9.44458
8.43	71.0649	2.90345	9.18150	8.93	79.7449	2.98831	9.44987
8.44	71.2336	2.90517	9.18695	8.94	79.9236	2.98998	9.45516
8.45	71.4025	2.90689	9.19239	8.95	80.1025	2.99166	9.46044
8.46	71.5716	2.90861	9.19783	8.96	80.2816	2.99333	9.46573
8.47	71.7409	2.91033	9.20326	8.97	80.4609	2.99500	9.47101
8.48	71.9104	2.91204	9.20869	8.98	80.6404	2.99666	9.47629
8.49	72.0801	2.91376	9.21412	8.99	80.8201	2.99833	9.48156
8.50	72.2500	2.91548	9.21954	**9.00**	81.0000	3.00000	9.48683
N	N²	√N	√10N	N	N²	√N	√10N

TABLE II *(continued)*

N	N²	√N	√10N	N	N²	√N	√10N
9.00	81.0000	3.00000	9.48683	**9.50**	90.2500	3.08221	9.74679
9.01	81.1801	3.00167	9.49210	9.51	90.4401	3.08383	9.75192
9.02	81.3604	3.00333	9.49737	9.52	90.6304	3.08545	9.75705
9.03	81.5409	3.00500	9.50263	9.53	90.8209	3.08707	9.76217
9.04	81.7216	3.00666	9.50789	9.54	91.0116	3.08869	9.76729
9.05	81.9025	3.00832	9.51315	9.55	91.2025	3.09031	9.77241
9.06	82.0836	3.00998	9.51840	9.56	91.3936	3.09192	9.77753
9.07	82.2649	3.01164	9.52365	9.57	91.5849	3.09354	9.78264
9.08	82.4464	3.01330	9.52890	9.58	91.7764	3.09516	9.78775
9.09	82.6281	3.01496	9.53415	9.59	91.9681	3.09677	9.79285
9.10	82.8100	3.01662	9.53939	**9.60**	92.1600	3.09839	9.79796
9.11	82.9921	3.01828	9.54463	9.61	92.3521	3.10000	9.80306
9.12	83.1744	3.01993	9.54987	9.62	92.5444	3.10161	9.80816
9.13	83.3569	3.02159	9.55510	9.63	92.7369	3.10322	9.81326
9.14	83.5396	3.02324	9.56033	9.64	92.9296	3.10483	9.81835
9.15	83.7225	3.02490	9.56556	9.65	93.1225	3.10644	9.82344
9.16	83.9056	3.02655	9.57079	9.66	93.3156	3.10805	9.82853
9.17	84.0889	3.02820	9.57601	9.67	93.5089	3.10966	9.83362
9.18	84.2724	3.02985	9.58123	9.68	93.7024	3.11127	9.83870
9.19	84.4561	3.03150	9.58645	9.69	93.8961	3.11288	9.84378
9.20	84.6400	3.03315	9.59166	**9.70**	94.0900	3.11448	9.84886
9.21	84.8241	3.03480	9.59687	9.71	94.2841	3.11609	9.85393
9.22	85.0084	3.03645	9.60208	9.72	94.4784	3.11769	9.85901
9.23	85.1929	3.03809	9.60729	9.73	94.6729	3.11929	9.86408
9.24	85.3776	3.03974	9.61249	9.74	94.8676	3.12090	9.86914
9.25	85.5625	3.04138	9.61769	9.75	95.0625	3.12250	9.87421
9.26	85.7476	3.04302	9.62289	9.76	95.2576	3.12410	9.87927
9.27	85.9329	3.04467	9.62808	9.77	95.4529	3.12570	9.88433
9.28	86.1184	3.04631	9.63328	9.78	95.6484	3.12730	9.88939
9.29	86.3041	3.04795	9.63846	9.79	95.8441	3.12890	9.89444
9.30	86.4900	3.04959	9.64365	**9.80**	96.0400	3.13050	9.89949
9.31	86.6761	3.05123	9.64883	9.81	96.2361	3.13209	9.90454
9.32	86.8624	3.05287	9.65401	9.82	96.4324	3.13369	9.90959
9.33	87.0489	3.05450	9.65919	9.83	96.6289	3.13528	9.91464
9.34	87.2356	3.05614	9.66437	9.84	96.8256	3.13688	9.91968
9.35	87.4225	3.05778	9.66954	9.85	97.0225	3.13847	9.92472
9.36	87.6096	3.05941	9.67471	9.86	97.2196	3.14006	9.92974
9.37	87.7969	3.06105	9.67988	9.87	97.4169	3.14166	9.93479
9.38	87.9844	3.06268	9.68504	9.88	97.6144	3.14325	9.93982
9.39	88.1721	3.06431	9.69020	9.89	97.8121	3.14484	9.94485
9.40	88.3600	3.06594	9.69536	**9.90**	98.0100	3.14643	9.94987
9.41	88.5481	3.06757	9.70052	9.91	98.2081	3.14802	9.95490
9.42	88.7364	3.06920	9.70567	9.92	98.4064	3.14960	9.95992
9.43	88.9249	3.07083	9.71082	9.93	98.6049	3.15119	9.96494
9.44	89.1136	3.07246	9.71597	9.94	98.8036	3.15278	9.96995
9.45	89.3025	3.07409	9.72111	9.95	99.0025	3.15436	9.97497
9.46	89.4916	3.07571	9.72625	9.96	99.2016	3.15595	9.97998
9.47	89.6809	3.07734	9.73139	9.97	99.4009	3.15753	9.98499
9.48	89.8704	3.07896	9.73653	9.98	99.6004	3.15911	9.98999
9.49	90.0601	3.08058	9.74166	9.99	99.8001	3.16070	9.99500
9.50	90.2500	3.08221	9.74679	**10.0**	100.000	3.16228	10.0000
N	N²	√N	√10N	N	N²	√N	√10N

From Paul G. Hoel, *Elementary Statistics*, 3rd ed., © 1971, John Wiley and Sons, Inc., New York, pp. 275-283.

TABLE III BINOMIAL PROBABILITIES

n	x	0.05	0.1	0.2	0.25	0.3	0.4	P 0.5	0.6	0.7	0.75	0.8	0.9	0.95
2	0	0.902	0.810	0.640	0.563	0.490	0.360	0.250	0.160	0.090	0.063	0.040	0.010	0.002
	1	0.095	0.180	0.320	0.375	0.420	0.480	0.500	0.480	0.420	0.375	0.320	0.180	0.095
	2	0.002	0.010	0.040	0.063	0.090	0.160	0.250	0.360	0.490	0.563	0.640	0.810	0.902
3	0	0.857	0.729	0.512	0.422	0.343	0.216	0.125	0.064	0.027	0.016	0.008	0.001	
	1	0.135	0.243	0.384	0.422	0.441	0.432	0.375	0.288	0.189	0.141	0.096	0.027	0.007
	2	0.007	0.027	0.096	0.141	0.189	0.288	0.375	0.432	0.441	0.422	0.384	0.243	0.135
	3		0.001	0.008	0.016	0.027	0.064	0.125	0.216	0.343	0.422	0.512	0.729	0.857
4	0	0.815	0.656	0.410	0.316	0.240	0.130	0.062	0.026	0.008	0.004	0.002		
	1	0.171	0.292	0.410	0.422	0.412	0.346	0.250	0.154	0.076	0.047	0.026	0.004	
	2	0.014	0.049	0.154	0.211	0.265	0.346	0.375	0.346	0.265	0.211	0.154	0.049	0.014
	3		0.004	0.026	0.047	0.076	0.154	0.250	0.346	0.412	0.422	0.410	0.292	0.171
	4			0.002	0.004	0.008	0.026	0.062	0.130	0.240	0.316	0.410	0.656	0.815
5	0	0.774	0.590	0.328	0.237	0.168	0.078	0.031	0.010	0.002	0.001			
	1	0.204	0.328	0.410	0.396	0.360	0.259	0.156	0.007	0.028	0.015	0.006		
	2	0.021	0.073	0.205	0.264	0.309	0.346	0.312	0.230	0.132	0.088	0.051	0.008	0.001
	3	0.001	0.008	0.051	0.088	0.132	0.230	0.312	0.346	0.309	0.274	0.205	0.073	0.021
	4			0.006	0.015	0.028	0.077	0.156	0.259	0.360	0.396	0.410	0.328	0.204
	5				0.001	0.002	0.010	0.031	0.078	0.168	0.237	0.328	0.590	0.774
6	0	0.735	0.531	0.262	0.178	0.118	0.047	0.016	0.004	0.001				
	1	0.232	0.354	0.393	0.356	0.303	0.187	0.094	0.037	0.010	0.004	0.002		
	2	0.031	0.098	0.246	0.297	0.324	0.311	0.234	0.138	0.060	0.033	0.015	0.001	
	3	0.002	0.015	0.082	0.132	0.185	0.276	0.132	0.276	0.185	0.132	0.082	0.015	0.002
	4		0.001	0.015	0.033	0.060	0.138	0.234	0.311	0.324	0.297	0.246	0.098	0.031
	5			0.002	0.004	0.010	0.037	0.094	0.187	0.303	0.356	0.393	0.354	0.232
	6					0.001	0.004	0.016	0.047	0.118	0.178	0.262	0.531	0.735
7	0	0.698	0.478	0.210	0.134	0.082	0.028	0.008	0.002					
	1	0.257	0.372	0.367	0.312	0.247	0.131	0.055	0.017	0.004	0.001			
	2	0.041	0.124	0.275	0.312	0.318	0.261	0.164	0.077	0.025	0.012	0.004		
	3	0.004	0.023	0.115	0.173	0.227	0.290	0.273	0.194	0.097	0.058	0.029	0.003	
	4		0.003	0.029	0.058	0.097	0.194	0.273	0.290	0.227	0.173	0.115	0.023	0.004
	5			0.004	0.012	0.025	0.077	0.164	0.261	0.318	0.312	0.275	0.124	0.041
	6				0.001	0.004	0.017	0.055	0.131	0.247	0.312	0.367	0.372	0.257
	7						0.002	0.008	0.028	0.082	0.134	0.210	0.478	0.698
8	0	0.663	0.430	0.168	0.100	0.058	0.017	0.004	0.001					
	1	0.279	0.383	0.336	0.267	0.198	0.090	0.031	0.008	0.001				
	2	0.051	0.149	0.294	0.312	0.296	0.209	0.109	0.041	0.010	0.004	0.001		
	3	0.005	0.033	0.147	0.208	0.254	0.279	0.219	0.124	0.047	0.023	0.009		
	4		0.005	0.046	0.087	0.136	0.232	0.273	0.232	0.136	0.087	0.046	0.005	
	5			0.009	0.023	0.047	0.124	0.219	0.279	0.254	0.208	0.147	0.033	0.005
	6			0.001	0.004	0.010	0.041	0.109	0.209	0.296	0.312	0.294	0.149	0.051
	7					0.001	0.008	0.031	0.090	0.198	0.267	0.336	0.383	0.279
	8						0.001	0.004	0.017	0.058	0.100	0.168	0.430	0.663

TABLE III BINOMIAL PROBABILITIES *(continued)*

n	x	0.05	0.1	0.2	0.25	0.3	0.4	0.5	0.6	0.7	0.75	0.8	0.9	0.95
9	0	0.630	0.387	0.134	0.075	0.040	0.010	0.002						
	1	0.299	0.387	0.302	0.225	0.156	0.060	0.018	0.004					
	2	0.063	0.172	0.302	0.300	0.267	0.161	0.070	0.021	0.004	0.001			
	3	0.008	0.045	0.176	0.234	0.267	0.251	0.164	0.074	0.021	0.009	0.003		
	4	0.001	0.007	0.066	0.117	0.172	0.251	0.246	0.167	0.074	0.039	0.017	0.001	
	5		0.001	0.017	0.039	0.074	0.167	0.246	0.251	0.172	0.117	0.066	0.007	0.001
	6			0.003	0.009	0.021	0.074	0.164	0.251	0.267	0.234	0.176	0.045	0.008
	7				0.001	0.004	0.021	0.070	0.161	0.267	0.300	0.302	0.172	0.063
	8						0.004	0.018	0.060	0.156	0.225	0.302	0.387	0.299
	9							0.002	0.010	0.040	0.075	0.134	0.387	0.630
10	0	0.599	0.349	0.107	0.056	0.028	0.006	0.001						
	1	0.315	0.387	0.268	0.188	0.121	0.040	0.010	0.002					
	2	0.075	0.194	0.302	0.282	0.233	0.121	0.044	0.011	0.001				
	3	0.010	0.057	0.201	0.250	0.267	0.215	0.117	0.042	0.009	0.003	0.001		
	4	0.001	0.011	0.088	0.146	0.200	0.251	0.205	0.111	0.037	0.016	0.006		
	5		0.001	0.026	0.058	0.103	0.201	0.246	0.201	0.103	0.058	0.026	0.001	
	6			0.006	0.016	0.037	0.111	0.205	0.251	0.200	0.146	0.088	0.011	0.001
	7			0.001	0.003	0.009	0.042	0.117	0.215	0.267	0.250	0.201	0.057	0.010
	8					0.001	0.011	0.044	0.121	0.233	0.282	0.302	0.194	0.075
	9						0.002	0.010	0.040	0.121	0.188	0.268	0.387	0.315
	10							0.001	0.006	0.028	0.056	0.107	0.349	0.599
11	0	0.569	0.314	0.086	0.042	0.020	0.004							
	1	0.329	0.384	0.236	0.155	0.093	0.027	0.005	0.001					
	2	0.087	0.213	0.295	0.258	0.200	0.089	0.027	0.005	0.001				
	3	0.014	0.071	0.221	0.258	0.257	0.177	0.081	0.023	0.004	0.001			
	4	0.001	0.016	0.111	0.172	0.220	0.236	0.161	0.070	0.017	0.006	0.002		
	5		0.002	0.039	0.080	0.132	0.221	0.226	0.147	0.057	0.027	0.010		
	6			0.010	0.027	0.057	0.147	0.226	0.221	0.132	0.080	0.039	0.002	
	7			0.002	0.006	0.017	0.070	0.161	0.236	0.220	0.172	0.111	0.016	0.001
	8				0.001	0.004	0.023	0.081	0.177	0.257	0.258	0.221	0.071	0.014
	9					0.001	0.005	0.027	0.089	0.200	0.258	0.295	0.213	0.087
	10						0.001	0.005	0.027	0.093	0.155	0.236	0.384	0.329
	11								0.004	0.020	0.042	0.086	0.314	0.569
12	0	0.540	0.282	0.069	0.032	0.014	0.002							
	1	0.341	0.377	0.206	0.127	0.071	0.017	0.003						
	2	0.099	0.230	0.283	0.232	0.168	0.064	0.016	0.002					
	3	0.017	0.085	0.236	0.258	0.240	0.142	0.054	0.012	0.001				
	4	0.002	0.021	0.133	0.194	0.231	0.213	0.121	0.042	0.008	0.002	0.001		
	5		0.004	0.053	0.103	0.158	0.227	0.193	0.101	0.029	0.012	0.003		
	6			0.016	0.040	0.079	0.177	0.226	0.177	0.079	0.040	0.016		
	7			0.003	0.012	0.029	0.101	0.193	0.227	0.158	0.103	0.053	0.004	
	8			0.001	0.002	0.008	0.042	0.121	0.213	0.231	0.194	0.133	0.021	0.002
	9					0.001	0.012	0.054	0.142	0.240	0.258	0.236	0.085	0.017
	10						0.002	0.016	0.064	0.168	0.232	0.283	0.230	0.099
	11							0.003	0.017	0.071	0.127	0.206	0.377	0.341
	12								0.002	0.014	0.032	0.069	0.282	0.540

TABLE III   BINOMIAL PROBABILITIES *(continued)*

n	x	0.05	0.1	0.2	0.25	0.3	0.4	0.5	0.6	0.7	0.75	0.8	0.9	0.95
13	0	0.513	0.254	0.055	0.024	0.010	0.001							
	1	0.351	0.367	0.179	0.103	0.054	0.011	0.002						
	2	0.111	0.245	0.268	0.206	0.139	0.045	0.010	0.001					
	3	0.021	0.100	0.246	0.252	0.218	0.111	0.035	0.006	0.001				
	4	0.003	0.028	0.154	0.210	0.234	0.184	0.087	0.024	0.003				
	5		0.006	0.069	0.126	0.180	0.221	0.157	0.066	0.014	0.005	0.001		
	6		0.001	0.023	0.056	0.103	0.197	0.209	0.131	0.044	0.019	0.006		
	7			0.006	0.019	0.044	0.131	0.209	0.197	0.103	0.056	0.023	0.001	
	8			0.001	0.005	0.014	0.066	0.157	0.221	0.180	0.126	0.069	0.006	
	9				0.001	0.003	0.024	0.087	0.184	0.234	0.210	0.154	0.028	0.003
	10					0.001	0.006	0.035	0.111	0.218	0.252	0.246	0.100	0.021
	11						0.001	0.010	0.045	0.139	0.206	0.268	0.245	0.111
	12							0.002	0.011	0.054	0.103	0.179	0.367	0.351
	13							0.001	0.010	0.024	0.055	0.254	0.513	
14	0	0.488	0.229	0.044	0.018	0.007	0.001							
	1	0.359	0.356	0.154	0.083	0.041	0.007	0.001						
	2	0.123	0.257	0.250	0.180	0.113	0.032	0.006	0.001					
	3	0.026	0.114	0.250	0.240	0.194	0.085	0.022	0.003					
	4	0.004	0.035	0.172	0.220	0.229	0.155	0.061	0.014	0.001				
	5		0.008	0.086	0.147	0.196	0.207	0.122	0.041	0.007	0.002			
	6		0.001	0.032	0.073	0.126	0.207	0.183	0.092	0.023	0.008	0.002		
	7			0.009	0.028	0.062	0.157	0.209	0.157	0.062	0.028	0.009		
	8			0.002	0.008	0.023	0.092	0.183	0.207	0.126	0.073	0.032	0.001	
	9				0.002	0.007	0.041	0.122	0.207	0.196	0.147	0.086	0.008	
	10					0.001	0.014	0.061	0.155	0.229	0.220	0.172	0.035	0.004
	11						0.003	0.022	0.085	0.194	0.240	0.250	0.114	0.026
	12						0.001	0.006	0.032	0.113	0.180	0.250	0.257	0.123
	13							0.001	0.007	0.041	0.083	0.154	0.356	0.359
	14								0.001	0.007	0.018	0.044	0.229	0.488
15	0	0.463	0.206	0.035	0.013	0.005								
	1	0.366	0.343	0.132	0.067	0.031	0.005							
	2	0.135	0.267	0.231	0.156	0.092	0.022	0.003						
	3	0.031	0.129	0.250	0.225	0.170	0.063	0.014	0.002					
	4	0.005	0.043	0.188	0.225	0.219	0.127	0.042	0.007	0.001				
	5	0.001	0.010	0.103	0.165	0.206	0.186	0.092	0.024	0.003	0.001			
	6		0.002	0.043	0.092	0.147	0.207	0.153	0.061	0.012	0.003	0.001		
	7			0.014	0.039	0.081	0.177	0.196	0.118	0.035	0.013	0.003		
	8			0.003	0.013	0.035	0.118	0.196	0.177	0.081	0.039	0.014		
	9			0.001	0.003	0.012	0.061	0.153	0.207	0.147	0.092	0.043	0.002	
	10				0.001	0.003	0.024	0.092	0.186	0.206	0.165	0.103	0.010	0.001
	11					0.001	0.007	0.042	0.127	0.219	0.225	0.188	0.043	0.005
	12						0.002	0.014	0.063	0.170	0.225	0.250	0.129	0.031
	13							0.003	0.022	0.092	0.156	0.231	0.267	0.135
	14								0.005	0.031	0.067	0.132	0.343	0.366
	15									0.005	0.013	0.035	0.206	0.463

TABLE IV   AREAS UNDER THE NORMAL CURVE

An entry in the table is the proportion under the entire curve which is between $z = 0$ and a positive value of z. Areas for negative values of z are obtained by symmetry.

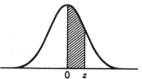

Second decimal place of z

z	.00	.01	.02	.03	.04	.05	.06	.07	.08	.09
0.0	.0000	.0040	.0080	.0120	.0160	.0199	.0239	.0279	.0319	.0359
0.1	.0398	.0438	.0478	.0517	.0557	.0596	.0636	.0675	.0714	.0753
0.2	.0793	.0832	.0871	.0910	.0948	.0987	.1026	.1064	.1103	.1141
0.3	.1179	.1217	.1255	.1293	.1331	.1368	.1406	.1443	.1480	.1517
0.4	.1554	.1591	.1628	.1664	.1700	.1736	.1772	.1808	.1844	.1879
0.5	.1915	.1950	.1985	.2019	.2054	.2088	.2123	.2157	.2190	.2224
0.6	.2257	.2291	.2324	.2357	.2389	.2422	.2454	.2486	.2517	.2549
0.7	.2580	.2611	.2642	.2673	.2703	.2734	.2764	.2794	.2823	.2852
0.8	.2881	.2910	.2939	.2967	.2995	.3023	.3051	.3078	.3106	.3133
0.9	.3159	.3186	.3212	.3238	.3264	.3289	.3315	.3340	.3365	.3389
1.0	.3413	.3438	.3461	.3485	.3508	.3531	.3554	.3577	.3599	.3621
1.1	.3643	.3665	.3686	.3708	.3729	.3749	.3770	.3790	.3810	.3830
1.2	.3849	.3869	.3888	.3907	.3925	.3944	.3962	.3980	.3997	.4015
1.3	.4032	.4049	.4066	.4082	.4099	.4115	.4131	.4147	.4162	.4177
1.4	.4192	.4207	.4222	.4236	.4251	.4265	.4279	.4292	.4306	.4319
1.5	.4332	.4345	.4357	.4370	.4382	.4394	.4406	.4418	.4429	.4441
1.6	.4452	.4463	.4474	.4484	.4495	.4505	.4515	.4525	.4535	.4545
1.7	.4554	.4564	.4573	.4582	.4591	.4599	.4608	.4616	.4625	.4633
1.8	.4641	.4649	.4656	.4664	.4671	.4678	.4686	.4693	.4699	.4706
1.9	.4713	.4719	.4726	.4732	.4738	.4744	.4750	.4756	.4761	.4767
2.0	.4772	.4778	.4783	.4788	.4793	.4798	.4803	.4808	.4812	.4817
2.1	.4821	.4826	.4830	.4834	.4838	.4842	.4846	.4850	.4854	.4857
2.2	.4861	.4864	.4868	.4871	.4875	.4878	.4881	.4884	.4887	.4890
2.3	.4893	.4896	.4898	.4901	.4904	.4906	.4909	.4911	.4913	.4916
2.4	.4918	.4920	.4922	.4925	.4927	.4929	.4931	.4932	.4934	.4936
2.5	.4938	.4940	.4941	.4943	.4945	.4946	.4948	.4949	.4951	.4952
2.6	.4953	.4955	.4956	.4957	.4959	.4960	.4961	.4962	.4963	.4964
2.7	.4965	.4966	.4967	.4968	.4969	.4970	.4971	.4972	.4973	.4974
2.8	.4974	.4975	.4976	.4977	.4977	.4978	.4979	.4979	.4980	.4981
2.9	.4981	.4982	.4982	.4983	.4984	.4984	.4985	.4985	.4986	.4986
3.0	.4987	.4987	.4987	.4988	.4988	.4989	.4989	.4989	.4990	.4990

From Paul G. Hoel, *Elementary Statistics*, 3rd ed., © 1971, John Wiley and Sons, Inc., New York, p. 287.

TABLE V   CRITICAL POINTS OF THE t DISTRIBUTION

The first column lists the number of degrees of freedom (ν). The headings of the other columns give probabilities (P) for t to exceed the entry value. Use symmetry for negative t values.

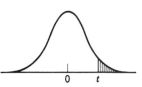

P df	.10	.05	.025	.01	.005
1	3.078	6.314	12.706	31.821	63.657
2	1.886	2.920	4.303	6.965	9.925
3	1.638	2.353	3.182	4.541	5.841
4	1.533	2.132	2.776	3.747	4.604
5	1.476	2.015	2.571	3.365	4.032
6	1.440	1.943	2.447	3.143	3.707
7	1.415	1.895	2.365	2.998	3.499
8	1.397	1.860	2.306	2.896	3.355
9	1.383	1.833	2.262	2.821	3.250
10	1.372	1.812	2.228	2.764	3.169
11	1.363	1.796	2.201	2.718	3.106
12	1.356	1.782	2.179	2.681	3.055
13	1.350	1.771	2.160	2.650	3.012
14	1.345	1.761	2.145	2.624	2.977
15	1.341	1.753	2.131	2.602	2.947
16	1.337	1.746	2.120	2.583	2.921
17	1.333	1.740	2.110	2.567	2.898
18	1.330	1.734	2.101	2.552	2.878
19	1.328	1.729	2.093	2.539	2.861
20	1.325	1.725	2.086	2.528	2.845
21	1.323	1.721	2.080	2.518	2.831
22	1.321	1.717	2.074	2.508	2.819
23	1.319	1.714	2.069	2.500	2.807
24	1.318	1.711	2.064	2.492	2.797
25	1.316	1.708	2.060	2.485	2.787
26	1.315	1.706	2.056	2.479	2.779
27	1.314	1.703	2.052	2.473	2.771
28	1.313	1.701	2.048	2.467	2.763
29	1.311	1.699	2.045	2.462	2.756
30	1.310	1.697	2.042	2.457	2.750
40	1.303	1.684	2.021	2.423	2.704
60	1.296	1.671	2.000	2.390	2.660
120	1.289	1.658	1.980	2.358	2.617
∞	1.282	1.645	1.960	2.326	2.576

From Paul G. Hoel, *Elementary Statistics*, 3rd ed., © 1971, John Wiley and Sons, Inc., New York, p. 288.

TABLE VI   CRITICAL POINTS OF THE *F* DISTRIBUTION

5% (Roman Type) and 1% (Boldface Type) Points for the Distribution of *F*

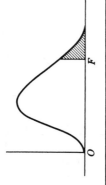

Each cell lists the 5% (Roman) point on the first line and the 1% (boldface) point on the second line.

Degrees of freedom for denominator (df_2)	\	1	2	3	4	5	6	7	8	9	10	11	12	14	16	20	24	30	40	50	75	100	200	500	∞
1	5%	161	200	216	225	230	234	237	239	241	242	243	244	245	246	248	249	250	251	252	253	253	254	254	254
1	1%	**4052**	**4999**	**5403**	**5625**	**5764**	**5859**	**5928**	**5981**	**6022**	**6056**	**6082**	**6106**	**6142**	**6169**	**6208**	**6234**	**6258**	**6286**	**6302**	**6323**	**6334**	**6352**	**6361**	**6366**
2	5%	18.51	19.00	19.16	19.25	19.30	19.33	19.36	19.37	19.38	19.39	19.40	19.41	19.42	19.43	19.44	19.45	19.46	19.47	19.47	19.48	19.49	19.49	19.50	19.50
2	1%	**98.49**	**99.01**	**99.17**	**99.25**	**99.30**	**99.33**	**99.34**	**99.36**	**99.38**	**99.40**	**99.41**	**99.42**	**99.43**	**99.44**	**99.45**	**99.46**	**99.47**	**99.48**	**99.48**	**99.49**	**99.49**	**99.49**	**99.50**	**99.50**
3	5%	10.13	9.55	9.28	9.12	9.01	8.94	8.88	8.84	8.81	8.78	8.76	8.74	8.71	8.69	8.66	8.64	8.62	8.60	8.58	8.57	8.56	8.54	8.54	8.53
3	1%	**34.12**	**30.81**	**29.46**	**28.71**	**28.24**	**27.91**	**27.67**	**27.49**	**27.34**	**27.23**	**27.13**	**27.05**	**26.92**	**26.83**	**26.69**	**26.60**	**26.50**	**26.41**	**26.30**	**26.27**	**26.23**	**26.18**	**26.14**	**26.12**
4	5%	7.71	6.94	6.59	6.39	6.26	6.16	6.09	6.04	6.00	5.96	5.93	5.91	5.87	5.84	5.80	5.77	5.74	5.71	5.70	5.68	5.66	5.65	5.64	5.63
4	1%	**21.20**	**18.00**	**16.69**	**15.98**	**15.52**	**15.21**	**14.98**	**14.80**	**14.66**	**14.54**	**14.45**	**14.37**	**14.24**	**14.15**	**14.02**	**13.93**	**13.83**	**13.74**	**13.69**	**13.61**	**13.57**	**13.52**	**13.48**	**13.46**
5	5%	6.61	5.79	5.41	5.19	5.05	4.95	4.88	4.82	4.78	4.74	4.70	4.68	4.64	4.60	4.56	4.53	4.50	4.46	4.44	4.42	4.40	4.38	4.37	4.36
5	1%	**16.26**	**13.27**	**12.06**	**11.39**	**10.97**	**10.67**	**10.45**	**10.27**	**10.15**	**10.05**	**9.96**	**9.89**	**9.77**	**9.68**	**9.55**	**9.47**	**9.38**	**9.29**	**9.24**	**9.17**	**9.13**	**9.07**	**9.04**	**9.02**
6	5%	5.99	5.14	4.76	4.53	4.39	4.28	4.21	4.15	4.10	4.06	4.03	4.00	3.96	3.92	3.87	3.84	3.81	3.77	3.75	3.72	3.71	3.69	3.68	3.67
6	1%	**13.74**	**10.92**	**9.78**	**9.15**	**8.75**	**8.47**	**8.26**	**8.10**	**7.98**	**7.87**	**7.79**	**7.72**	**7.60**	**7.52**	**7.39**	**7.31**	**7.23**	**7.14**	**7.09**	**7.02**	**6.99**	**6.94**	**6.90**	**6.88**
7	5%	5.59	4.74	4.35	4.12	3.97	3.87	3.79	3.73	3.68	3.63	3.60	3.57	3.52	3.49	3.44	3.41	3.38	3.34	3.32	3.29	3.28	3.25	3.24	3.23
7	1%	**12.25**	**9.55**	**8.45**	**7.85**	**7.46**	**7.19**	**7.00**	**6.84**	**6.71**	**6.62**	**6.54**	**6.47**	**6.35**	**6.27**	**6.15**	**6.07**	**5.98**	**5.90**	**5.85**	**5.78**	**5.75**	**5.70**	**5.67**	**5.65**
8	5%	5.32	4.46	4.07	3.84	3.69	3.58	3.50	3.44	3.39	3.34	3.31	3.28	3.23	3.20	3.15	3.12	3.08	3.05	3.03	3.00	2.98	2.96	2.94	2.93
8	1%	**11.26**	**8.65**	**7.59**	**7.01**	**6.63**	**6.37**	**6.19**	**6.03**	**5.91**	**5.82**	**5.74**	**5.67**	**5.56**	**5.48**	**5.36**	**5.28**	**5.20**	**5.11**	**5.06**	**5.00**	**4.96**	**4.91**	**4.88**	**4.86**
9	5%	5.12	4.26	3.86	3.63	3.48	3.37	3.29	3.23	3.18	3.13	3.10	3.07	3.02	2.98	2.93	2.90	2.86	2.82	2.80	2.77	2.76	2.73	2.72	2.71
9	1%	**10.56**	**8.02**	**6.99**	**6.42**	**6.06**	**5.80**	**5.62**	**5.47**	**5.35**	**5.26**	**5.18**	**5.11**	**5.00**	**4.92**	**4.80**	**4.73**	**4.64**	**4.56**	**4.51**	**4.45**	**4.41**	**4.36**	**4.33**	**4.31**

Degrees of freedom for numerator (df_1)

TABLE VI *(continued)*

Degrees of freedom for numerator (df_1)

Degrees of freedom for denominator (df_2)	1	2	3	4	5	6	7	8	9	10	11	12	14	16	20	24	30	40	50	75	100	200	500	∞
10	4.96 / 10.04	4.10 / 7.56	3.71 / 6.55	3.48 / 5.99	3.33 / 5.64	3.22 / 5.39	3.14 / 5.21	3.07 / 5.06	3.02 / 4.95	2.97 / 4.85	2.94 / 4.78	2.91 / 4.71	2.86 / 4.60	2.82 / 4.52	2.77 / 4.41	2.74 / 4.33	2.70 / 4.25	2.67 / 4.17	2.64 / 4.12	2.61 / 4.05	2.59 / 4.01	2.56 / 3.96	2.55 / 3.93	2.54 / 3.91
11	4.84 / 9.65	3.98 / 7.20	3.59 / 6.22	3.36 / 5.67	3.20 / 5.32	3.09 / 5.07	3.01 / 4.88	2.95 / 4.74	2.90 / 4.63	2.86 / 4.54	2.82 / 4.46	2.79 / 4.40	2.74 / 4.29	2.70 / 4.21	2.65 / 4.10	2.61 / 4.02	2.57 / 3.94	2.53 / 3.86	2.50 / 3.80	2.47 / 3.74	2.45 / 3.70	2.42 / 3.66	2.41 / 3.62	2.40 / 3.60
12	4.75 / 9.33	3.88 / 6.93	3.49 / 5.95	3.26 / 5.41	3.11 / 5.06	3.00 / 4.82	2.92 / 4.65	2.85 / 4.50	2.80 / 4.39	2.76 / 4.30	2.72 / 4.22	2.69 / 4.16	2.64 / 4.05	2.60 / 3.98	2.54 / 3.86	2.50 / 3.78	2.46 / 3.70	2.42 / 3.61	2.40 / 3.56	2.36 / 3.49	2.35 / 3.46	2.32 / 3.41	2.31 / 3.38	2.30 / 3.36
13	4.67 / 9.07	3.80 / 6.70	3.41 / 5.74	3.18 / 5.20	3.02 / 4.86	2.92 / 4.62	2.84 / 4.44	2.77 / 4.30	2.72 / 4.19	2.67 / 4.10	2.63 / 4.02	2.60 / 3.96	2.55 / 3.85	2.51 / 3.78	2.46 / 3.67	2.42 / 3.59	2.38 / 3.51	2.34 / 3.42	2.32 / 3.37	2.28 / 3.30	2.26 / 3.27	2.24 / 3.21	2.22 / 3.18	2.21 / 3.16
14	4.60 / 8.86	3.74 / 6.51	3.34 / 5.56	3.11 / 5.03	2.96 / 4.69	2.85 / 4.46	2.77 / 4.28	2.70 / 4.14	2.65 / 4.03	2.60 / 3.94	2.56 / 3.86	2.53 / 3.80	2.48 / 3.70	2.44 / 3.62	2.39 / 3.51	2.35 / 3.43	2.31 / 3.34	2.27 / 3.26	2.24 / 3.21	2.21 / 3.14	2.19 / 3.11	2.16 / 3.06	2.14 / 3.02	2.13 / 3.00
15	4.54 / 8.68	3.68 / 6.36	3.29 / 5.42	3.06 / 4.89	2.90 / 4.56	2.79 / 4.32	2.70 / 4.14	2.64 / 4.00	2.59 / 3.89	2.55 / 3.80	2.51 / 3.73	2.48 / 3.67	2.43 / 3.56	2.39 / 3.48	2.33 / 3.36	2.29 / 3.29	2.25 / 3.20	2.21 / 3.12	2.18 / 3.07	2.15 / 3.00	2.12 / 2.97	2.10 / 2.92	2.08 / 2.89	2.07 / 2.87
16	4.49 / 8.53	3.63 / 6.23	3.24 / 5.29	3.01 / 4.77	2.85 / 4.44	2.74 / 4.20	2.66 / 4.03	2.59 / 3.89	2.54 / 3.78	2.49 / 3.69	2.45 / 3.61	2.42 / 3.55	2.37 / 3.45	2.33 / 3.37	2.28 / 3.25	2.24 / 3.18	2.20 / 3.10	2.16 / 3.01	2.13 / 2.96	2.09 / 2.89	2.07 / 2.86	2.04 / 2.80	2.02 / 2.77	2.01 / 2.75
17	4.45 / 8.40	3.59 / 6.11	3.20 / 5.18	2.96 / 4.67	2.81 / 4.34	2.70 / 4.10	2.62 / 3.93	2.55 / 3.79	2.50 / 3.68	2.45 / 3.59	2.41 / 3.52	2.38 / 3.45	2.33 / 3.35	2.29 / 3.27	2.23 / 3.16	2.19 / 3.08	2.15 / 3.00	2.11 / 2.92	2.08 / 2.86	2.04 / 2.79	2.02 / 2.76	1.99 / 2.70	1.97 / 2.67	1.96 / 2.65
18	4.41 / 8.28	3.55 / 6.01	3.16 / 5.09	2.93 / 4.58	2.77 / 4.25	2.66 / 4.01	2.58 / 3.85	2.51 / 3.71	2.46 / 3.60	2.41 / 3.51	2.37 / 3.44	2.34 / 3.37	2.29 / 3.27	2.25 / 3.19	2.19 / 3.07	2.15 / 3.00	2.11 / 2.91	2.07 / 2.83	2.04 / 2.78	2.00 / 2.71	1.98 / 2.68	1.95 / 2.62	1.93 / 2.59	1.92 / 2.57
19	4.38 / 8.18	3.52 / 5.93	3.13 / 5.01	2.90 / 4.50	2.74 / 4.17	2.63 / 3.94	2.55 / 3.77	2.48 / 3.63	2.43 / 3.52	2.38 / 3.43	2.34 / 3.36	2.31 / 3.30	2.26 / 3.19	2.21 / 3.12	2.15 / 3.00	2.11 / 2.92	2.07 / 2.84	2.02 / 2.76	2.00 / 2.70	1.96 / 2.63	1.94 / 2.60	1.91 / 2.54	1.90 / 2.51	1.88 / 2.49
20	4.35 / 8.10	3.49 / 5.85	3.10 / 4.94	2.87 / 4.43	2.71 / 4.10	2.60 / 3.87	2.52 / 3.71	2.45 / 3.56	2.40 / 3.45	2.35 / 3.37	2.31 / 3.30	2.28 / 3.23	2.23 / 3.13	2.18 / 3.05	2.12 / 2.94	2.08 / 2.86	2.04 / 2.77	1.99 / 2.69	1.96 / 2.63	1.92 / 2.56	1.90 / 2.53	1.87 / 2.47	1.85 / 2.44	1.84 / 2.42
21	4.32 / 8.02	3.47 / 5.78	3.07 / 4.87	2.84 / 4.37	2.68 / 4.04	2.57 / 3.81	2.49 / 3.65	2.42 / 3.51	2.37 / 3.40	2.32 / 3.31	2.28 / 3.24	2.25 / 3.17	2.20 / 3.07	2.15 / 2.99	2.09 / 2.88	2.05 / 2.80	2.00 / 2.72	1.96 / 2.63	1.93 / 2.58	1.89 / 2.51	1.87 / 2.47	1.84 / 2.42	1.82 / 2.38	1.81 / 2.36
22	4.30 / 7.94	3.44 / 5.72	3.05 / 4.82	2.82 / 4.31	2.66 / 3.99	2.55 / 3.76	2.47 / 3.59	2.40 / 3.45	2.35 / 3.35	2.30 / 3.26	2.26 / 3.18	2.23 / 3.12	2.18 / 3.02	2.13 / 2.94	2.07 / 2.83	2.03 / 2.75	1.98 / 2.67	1.93 / 2.58	1.91 / 2.53	1.87 / 2.46	1.84 / 2.42	1.81 / 2.37	1.80 / 2.33	1.78 / 2.31
23	4.28 / 7.88	3.42 / 5.66	3.03 / 4.76	2.80 / 4.26	2.64 / 3.94	2.53 / 3.71	2.45 / 3.54	2.38 / 3.41	2.32 / 3.30	2.28 / 3.21	2.24 / 3.14	2.20 / 3.07	2.14 / 2.97	2.10 / 2.89	2.04 / 2.78	2.00 / 2.70	1.96 / 2.62	1.91 / 2.53	1.88 / 2.48	1.84 / 2.41	1.82 / 2.37	1.79 / 2.32	1.77 / 2.28	1.76 / 2.26

24	4.26 / 7.82	3.40 / 5.61	3.01 / 4.72	2.78 / 4.22	2.62 / 3.90	2.51 / 3.67	2.43 / 3.50	2.36 / 3.36	2.30 / 3.25	2.26 / 3.17	2.22 / 3.09	2.18 / 3.03	2.13 / 2.93	2.09 / 2.85	2.02 / 2.74	1.98 / 2.66	1.94 / 2.58	1.89 / 2.49	1.86 / 2.44	1.82 / 2.36	1.80 / 2.33	1.76 / 2.27	1.74 / 2.23	1.73 / 2.21
25	4.24 / 7.77	3.38 / 5.57	2.99 / 4.68	2.76 / 4.18	2.60 / 3.86	2.49 / 3.63	2.41 / 3.46	2.34 / 3.32	2.28 / 3.21	2.24 / 3.13	2.20 / 3.05	2.16 / 2.99	2.11 / 2.89	2.06 / 2.81	2.00 / 2.70	1.96 / 2.62	1.92 / 2.54	1.87 / 2.45	1.84 / 2.40	1.80 / 2.32	1.77 / 2.29	1.74 / 2.23	1.72 / 2.19	1.71 / 2.17
26	4.22 / 7.72	3.37 / 5.53	2.89 / 4.64	2.74 / 4.14	2.59 / 3.82	2.47 / 3.59	2.39 / 3.42	2.32 / 3.29	2.27 / 3.17	2.22 / 3.09	2.18 / 3.02	2.15 / 2.96	2.10 / 2.86	2.05 / 2.77	1.99 / 2.66	1.95 / 2.58	1.90 / 2.50	1.85 / 2.41	1.82 / 2.36	1.78 / 2.28	1.76 / 2.25	1.72 / 2.19	1.70 / 2.15	1.69 / 2.13
27	4.21 / 7.68	3.35 / 5.49	2.96 / 4.60	2.73 / 4.11	2.57 / 3.79	2.46 / 3.56	2.37 / 3.39	2.30 / 3.26	2.25 / 3.14	2.20 / 3.06	2.16 / 2.98	2.13 / 2.93	2.08 / 2.83	2.03 / 2.74	1.97 / 2.63	1.93 / 2.55	1.88 / 2.47	1.84 / 2.38	1.80 / 2.33	1.76 / 2.25	1.74 / 2.21	1.71 / 2.16	1.68 / 2.12	1.67 / 2.10
28	4.20 / 7.64	3.34 / 5.45	2.95 / 4.57	2.71 / 4.07	2.56 / 3.76	2.44 / 3.53	2.36 / 3.36	2.29 / 3.23	3.24 / 3.11	2.19 / 3.03	2.15 / 2.95	2.12 / 2.90	2.06 / 2.80	2.02 / 2.71	1.96 / 2.60	1.91 / 2.52	1.87 / 2.44	1.81 / 2.35	1.78 / 2.30	1.75 / 2.22	1.72 / 2.18	1.69 / 2.13	1.67 / 2.09	1.65 / 2.06
29	4.18 / 7.60	3.33 / 5.52	2.93 / 4.54	2.70 / 4.04	2.54 / 3.73	2.43 / 3.50	2.35 / 3.33	2.28 / 3.20	2.22 / 3.08	2.18 / 3.00	2.14 / 2.92	2.10 / 2.87	2.05 / 2.77	2.00 / 2.68	1.94 / 2.57	1.90 / 2.49	1.85 / 2.41	1.80 / 2.32	1.77 / 2.27	1.73 / 2.19	1.71 / 2.15	1.68 / 2.10	1.65 / 2.06	1.64 / 2.03
30	4.17 / 7.56	3.32 / 5.39	2.92 / 4.51	2.69 / 4.02	2.53 / 3.70	2.42 / 3.47	2.34 / 3.30	2.27 / 3.17	2.21 / 3.06	2.16 / 2.98	2.12 / 2.90	2.09 / 2.84	2.04 / 2.74	1.99 / 2.66	1.93 / 2.55	1.89 / 2.47	1.84 / 2.38	1.79 / 2.29	1.76 / 2.24	1.72 / 2.16	1.69 / 2.13	1.66 / 2.07	1.64 / 2.03	1.62 / 2.01
32	4.15 / 7.50	3.30 / 5.34	2.90 / 4.46	2.67 / 3.97	2.51 / 3.66	2.40 / 3.42	2.32 / 3.25	2.25 / 3.12	2.19 / 3.01	2.14 / 2.94	2.10 / 2.86	2.07 / 2.80	2.02 / 2.70	1.97 / 2.62	1.91 / 2.51	1.86 / 2.42	1.82 / 2.34	1.76 / 2.25	1.74 / 2.20	1.69 / 2.12	1.67 / 2.08	1.64 / 2.02	1.61 / 1.98	1.59 / 1.96
34	4.13 / 7.44	3.28 / 5.29	2.88 / 4.42	2.65 / 3.93	2.49 / 3.61	2.38 / 3.38	2.30 / 3.21	2.23 / 3.08	2.17 / 2.97	2.12 / 2.89	2.08 / 2.82	2.05 / 2.76	2.00 / 2.66	1.95 / 2.58	1.89 / 2.47	1.84 / 2.38	1.80 / 2.30	1.74 / 2.21	1.71 / 2.15	1.67 / 2.08	1.64 / 2.04	1.61 / 1.98	1.59 / 1.94	1.57 / 1.91
36	4.11 / 7.39	3.26 / 5.25	2.86 / 4.38	2.63 / 3.89	2.48 / 3.58	2.36 / 3.35	2.28 / 3.18	2.21 / 3.04	2.15 / 2.94	2.10 / 2.86	2.06 / 2.78	2.03 / 2.72	1.89 / 2.62	1.93 / 2.54	1.87 / 2.43	1.82 / 2.35	1.78 / 2.26	1.72 / 2.17	1.69 / 2.12	1.65 / 2.04	1.62 / 2.00	1.59 / 1.94	1.56 / 1.90	1.55 / 1.87
38	4.10 / 7.35	3.25 / 5.21	2.85 / 4.34	2.62 / 3.86	2.46 / 3.54	2.35 / 3.32	2.26 / 3.15	2.19 / 3.02	2.14 / 2.91	2.09 / 2.82	2.05 / 2.75	2.02 / 2.69	1.96 / 2.59	1.92 / 2.51	1.85 / 2.40	1.80 / 2.32	1.76 / 2.22	1.71 / 2.14	1.67 / 2.08	1.63 / 2.00	1.60 / 1.97	1.57 / 1.90	1.54 / 1.86	1.53 / 1.84
40	4.08 / 7.31	3.23 / 5.18	2.84 / 4.31	2.61 / 3.83	2.45 / 3.51	2.34 / 3.29	2.25 / 3.12	2.18 / 2.99	2.12 / 2.88	2.07 / 2.80	2.04 / 2.73	2.00 / 2.66	1.95 / 2.56	1.90 / 2.49	1.84 / 2.37	1.79 / 2.29	1.74 / 2.20	1.69 / 2.11	1.66 / 2.05	1.61 / 1.97	1.59 / 1.94	1.55 / 1.88	1.53 / 1.84	1.51 / 1.81
42	4.07 / 7.27	3.22 / 5.15	2.83 / 4.29	2.59 / 3.80	2.44 / 3.49	2.32 / 3.26	2.24 / 3.10	2.17 / 2.96	2.11 / 2.86	2.06 / 2.77	2.02 / 2.70	1.99 / 2.64	1.94 / 2.54	1.89 / 2.46	1.82 / 2.35	1.78 / 2.26	1.73 / 2.17	1.68 / 2.08	1.64 / 2.02	1.60 / 1.94	1.57 / 1.91	1.54 / 1.85	1.51 / 1.80	1.49 / 1.78
44	4.06 / 7.24	3.21 / 5.12	2.82 / 4.26	2.58 / 3.78	2.43 / 3.46	2.31 / 3.24	2.23 / 3.07	2.16 / 2.94	2.10 / 2.84	2.05 / 2.75	2.01 / 2.68	1.98 / 2.62	1.92 / 2.52	1.88 / 2.44	1.81 / 2.32	1.76 / 2.24	1.72 / 2.15	1.66 / 2.06	1.63 / 2.00	1.58 / 1.92	1.56 / 1.88	1.52 / 1.82	1.50 / 1.78	1.48 / 1.75
46	4.05 / 7.21	3.20 / 5.10	2.81 / 4.24	2.57 / 3.76	2.42 / 3.44	2.30 / 3.22	2.22 / 3.05	2.14 / 2.92	2.09 / 2.82	2.04 / 2.73	2.00 / 2.66	1.97 / 2.60	1.91 / 2.50	1.87 / 2.42	1.80 / 2.30	1.75 / 2.22	1.71 / 2.13	1.65 / 2.04	1.62 / 1.98	1.57 / 1.90	1.54 / 1.86	1.51 / 1.80	1.48 / 1.76	1.46 / 1.72
48	4.04 / 7.19	3.19 / 5.08	2.80 / 4.22	2.56 / 3.74	2.41 / 3.42	2.30 / 3.20	2.21 / 3.04	2.14 / 2.90	2.08 / 2.80	2.03 / 2.71	1.99 / 2.64	1.96 / 2.58	1.90 / 2.48	1.86 / 2.40	1.79 / 2.28	1.74 / 2.20	1.70 / 2.11	1.64 / 2.02	1.61 / 1.96	1.56 / 1.88	1.53 / 1.84	1.50 / 1.78	1.47 / 1.73	1.45 / 1.70

TABLE VI (continued)

Degrees of freedom for numerator (df_1). Each cell shows the upper value (roman) and lower value (**bold**).

Degrees of freedom for denominator (df_2)	1	2	3	4	5	6	7	8	9	10	11	12	14	16	20	24	30	40	50	75	100	200	500	∞
50	4.03 / **7.17**	3.18 / **5.06**	2.79 / **4.20**	2.56 / **3.72**	2.40 / **3.41**	2.29 / **3.18**	2.20 / **3.02**	2.13 / **2.88**	2.07 / **2.78**	2.02 / **2.70**	1.98 / **2.62**	1.95 / **2.56**	1.90 / **2.46**	1.85 / **2.39**	1.78 / **2.26**	1.74 / **2.18**	1.69 / **2.10**	1.63 / **2.00**	1.60 / **1.94**	1.55 / **1.86**	1.52 / **1.82**	1.48 / **1.76**	1.46 / **1.71**	1.44 / **1.68**
55	4.02 / **7.12**	3.17 / **5.01**	2.78 / **4.16**	2.54 / **3.68**	2.38 / **3.37**	2.27 / **3.15**	2.18 / **2.98**	2.11 / **2.85**	2.05 / **2.75**	2.00 / **2.66**	1.97 / **2.59**	1.93 / **2.53**	1.88 / **2.43**	1.83 / **2.35**	1.76 / **2.23**	1.72 / **2.15**	1.67 / **2.06**	1.61 / **1.96**	1.58 / **1.90**	1.52 / **1.82**	1.50 / **1.78**	1.46 / **1.71**	1.43 / **1.66**	1.41 / **1.64**
60	4.00 / **7.08**	3.15 / **4.98**	2.76 / **4.13**	2.52 / **3.65**	2.37 / **3.34**	2.25 / **3.12**	2.17 / **2.95**	2.10 / **2.82**	2.04 / **2.72**	1.99 / **2.63**	1.95 / **2.56**	1.92 / **2.50**	1.86 / **2.40**	1.81 / **2.32**	1.75 / **2.20**	1.70 / **2.12**	1.65 / **2.03**	1.59 / **1.93**	1.56 / **1.87**	1.50 / **1.79**	1.48 / **1.74**	1.44 / **1.68**	1.41 / **1.63**	1.39 / **1.60**
65	3.99 / **7.04**	3.14 / **4.95**	2.75 / **4.10**	2.51 / **3.62**	2.36 / **3.31**	2.24 / **3.09**	2.15 / **2.93**	2.08 / **2.79**	2.02 / **2.70**	1.98 / **2.61**	1.94 / **2.54**	1.90 / **2.47**	1.85 / **2.37**	1.80 / **2.30**	1.73 / **2.18**	1.68 / **2.09**	1.63 / **2.00**	1.57 / **1.90**	1.54 / **1.84**	1.49 / **1.76**	1.46 / **1.71**	1.42 / **1.64**	1.39 / **1.60**	1.37 / **1.56**
70	3.98 / **7.01**	3.13 / **4.92**	2.74 / **4.08**	2.50 / **3.60**	2.35 / **3.29**	2.23 / **3.07**	2.14 / **2.91**	2.07 / **2.77**	2.01 / **2.67**	1.97 / **2.59**	1.93 / **2.51**	1.89 / **2.45**	1.84 / **2.35**	1.79 / **2.28**	1.72 / **2.15**	1.67 / **2.07**	1.62 / **1.98**	1.56 / **1.88**	1.53 / **1.82**	1.47 / **1.74**	1.45 / **1.69**	1.40 / **1.63**	1.37 / **1.56**	1.35 / **1.53**
80	3.96 / **6.96**	3.11 / **4.88**	2.72 / **4.04**	2.48 / **3.56**	2.33 / **3.25**	2.21 / **3.04**	2.12 / **2.87**	2.05 / **2.74**	1.99 / **2.64**	1.95 / **2.55**	1.91 / **2.48**	1.88 / **2.41**	1.82 / **2.32**	1.77 / **2.24**	1.70 / **2.11**	1.65 / **2.03**	1.60 / **1.94**	1.54 / **1.84**	1.51 / **1.78**	1.45 / **1.70**	1.42 / **1.65**	1.38 / **1.57**	1.35 / **1.52**	1.32 / **1.49**
100	3.94 / **6.90**	3.09 / **4.82**	2.70 / **3.98**	2.46 / **3.51**	2.30 / **3.20**	2.19 / **2.99**	2.10 / **2.82**	2.03 / **2.69**	1.97 / **2.59**	1.92 / **2.51**	1.88 / **2.43**	1.85 / **2.36**	1.79 / **2.26**	1.75 / **2.19**	1.68 / **2.06**	1.63 / **1.98**	1.57 / **1.89**	1.51 / **1.79**	1.48 / **1.73**	1.42 / **1.64**	1.39 / **1.59**	1.34 / **1.51**	1.30 / **1.46**	1.28 / **1.43**
125	3.92 / **6.84**	3.07 / **4.78**	2.68 / **3.94**	2.44 / **3.47**	2.29 / **3.17**	2.17 / **2.95**	2.08 / **2.79**	2.01 / **2.65**	1.95 / **2.56**	1.90 / **2.47**	1.86 / **2.40**	1.83 / **2.33**	1.77 / **2.23**	1.72 / **2.15**	1.65 / **2.03**	1.60 / **1.94**	1.55 / **1.85**	1.49 / **1.75**	1.45 / **1.68**	1.39 / **1.59**	1.36 / **1.54**	1.31 / **1.46**	1.27 / **1.40**	1.25 / **1.37**
150	3.91 / **6.81**	3.06 / **4.75**	2.67 / **3.91**	2.43 / **3.44**	2.27 / **3.13**	2.16 / **2.92**	2.07 / **2.76**	2.00 / **2.62**	1.94 / **2.53**	1.89 / **2.44**	1.85 / **2.37**	1.82 / **2.30**	1.76 / **2.20**	1.71 / **2.12**	1.64 / **2.00**	1.59 / **1.91**	1.54 / **1.83**	1.47 / **1.72**	1.44 / **1.66**	1.37 / **1.56**	1.34 / **1.51**	1.29 / **1.43**	1.25 / **1.37**	1.22 / **1.33**
200	3.89 / **6.76**	3.04 / **4.71**	2.65 / **3.88**	2.41 / **3.41**	2.26 / **3.11**	2.14 / **2.90**	2.05 / **2.73**	1.98 / **2.60**	1.92 / **2.50**	1.87 / **2.41**	1.83 / **2.34**	1.80 / **2.28**	1.74 / **2.17**	1.69 / **2.09**	1.62 / **1.97**	1.57 / **1.88**	1.52 / **1.79**	1.45 / **1.69**	1.42 / **1.62**	1.35 / **1.53**	1.32 / **1.48**	1.26 / **1.39**	1.22 / **1.33**	1.19 / **1.28**
400	3.86 / **6.70**	3.02 / **4.66**	2.62 / **3.83**	2.39 / **3.36**	2.23 / **3.06**	2.12 / **2.85**	2.03 / **2.69**	1.96 / **2.55**	1.90 / **2.46**	1.85 / **2.37**	1.81 / **2.29**	1.78 / **2.23**	1.72 / **2.12**	1.67 / **2.04**	1.60 / **1.92**	1.54 / **1.84**	1.49 / **1.74**	1.42 / **1.64**	1.38 / **1.57**	1.32 / **1.47**	1.28 / **1.42**	1.22 / **1.32**	1.16 / **1.24**	1.13 / **1.19**
1000	3.85 / **6.66**	3.00 / **4.62**	2.61 / **3.80**	2.38 / **3.34**	2.22 / **3.04**	2.10 / **2.82**	2.02 / **2.66**	1.95 / **2.53**	1.89 / **2.43**	1.84 / **2.34**	1.80 / **2.26**	1.76 / **2.20**	1.70 / **2.09**	1.65 / **2.01**	1.58 / **1.89**	1.53 / **1.81**	1.47 / **1.71**	1.41 / **1.61**	1.36 / **1.54**	1.30 / **1.44**	1.26 / **1.38**	1.19 / **1.28**	1.13 / **1.19**	1.08 / **1.11**
∞	3.84 / **6.64**	2.99 / **4.60**	2.60 / **3.78**	2.37 / **3.32**	2.21 / **3.02**	2.09 / **2.80**	2.01 / **2.64**	1.94 / **2.51**	1.88 / **2.41**	1.83 / **2.32**	1.79 / **2.24**	1.75 / **2.18**	1.69 / **2.07**	1.64 / **1.99**	1.57 / **1.87**	1.52 / **1.79**	1.46 / **1.69**	1.40 / **1.59**	1.35 / **1.52**	1.28 / **1.41**	1.24 / **1.36**	1.17 / **1.25**	1.11 / **1.15**	1.00 / **1.00**

TABLE VII   CRITICAL VALUES OF r FOR TESTING $\rho = 0$

For a two-sided test α is twice the value listed at the heading of a column of critical r values; hence for $\alpha = 0.05$ choose the 0.025 column.

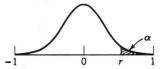

n \ α	0.05	0.025	0.010	0.005
5	0.805	0.878	0.934	0.959
6	0.729	0.811	0.882	0.917
7	0.669	0.754	0.833	0.875
8	0.621	0.707	0.789	0.834
9	0.582	0.666	0.750	0.798
10	0.549	0.632	0.716	0.765
11	0.521	0.602	0.685	0.735
12	0.497	0.576	0.658	0.708
13	0.476	0.553	0.634	0.684
14	0.457	0.532	0.612	0.661
15	0.441	0.514	0.592	0.641
16	0.426	0.497	0.574	0.623

n \ α	0.05	0.025	0.010	0.005
17	0.412	0.482	0.558	0.606
18	0.400	0.468	0.542	0.590
19	0.389	0.456	0.528	0.575
20	0.378	0.444	0.516	0.561
25	0.337	0.396	0.462	0.505
30	0.306	0.361	0.423	0.463
40	0.264	0.312	0.366	0.402
50	0.235	0.279	0.328	0.361
60	0.214	0.254	0.300	0.330
80	0.185	0.220	0.260	0.286
100	0.165	0.196	0.232	0.256

Tables VI and VII are from Paul G. Hoel, *Elementary Statistics*, 3rd ed., © 1971, John Wiley and Sons, Inc., New York, pp. 289, 292-294.

TABLE VIII CRITICAL POINTS OF THE χ^2 DISTRIBUTION

The first column lists the number of degrees of freedom. The headings of the other columns give probabilities (P) for χ^2 to exceed the entry value.

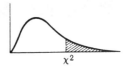

df	0.050	0.025	0.010	0.005
1	3.84146	5.02389	6.63490	7.87944
2	5.99147	7.37776	9.21034	10.5966
3	7.81473	9.34840	11.3449	12.8381
4	9.48773	11.1433	13.2767	14.8602
5	11.0705	12.8325	15.0863	16.7496
6	12.5916	14.4494	16.8119	18.5476
7	14.0671	16.0128	18.4753	20.2777
8	15.5073	17.5346	20.0902	21.9550
9	16.9190	19.0228	21.6660	23.5893
10	18.3070	20.4831	23.2093	25.1882
11	19.6751	21.9200	24.7250	26.7569
12	21.0261	23.3367	26.2170	28.2995
13	22.3621	24.7356	27.6883	29.8194
14	23.6848	26.1190	29.1413	31.3193
15	24.9958	27.4884	30.5779	32.8013
16	26.2962	28.8454	31.9999	34.2672
17	27.5871	30.1910	33.4087	35.7185
18	28.8693	31.5264	34.8053	37.1564
19	30.1435	32.8523	36.1908	38.5822
20	31.4104	34.1696	37.5662	39.9968
21	32.6705	35.4789	38.9321	41.4010
22	33.9244	36.7807	40.2894	42.7956
23	35.1725	38.0757	41.6384	44.1813
24	36.4151	39.3641	42.9798	45.5585
25	37.6525	40.6465	44.3141	46.9278
26	38.8852	41.9232	45.6417	48.2899
27	40.1133	43.1944	46.9630	49.6449
28	41.3372	44.4607	48.2782	50.9933
29	42.5569	45.7222	49.5879	52.3356
30	43.7729	46.9792	50.8922	53.6720
40	55.7585	59.3417	63.6907	66.7659
50	67.5048	71.4202	76.1539	79.4900
60	79.0819	83.2976	88.3794	91.9517
70	90.5312	95.0231	100.425	104.215
80	101.879	106.629	112.329	116.321
90	113.145	118.136	124.116	128.299
100	124.342	129.561	135.807	140.169

Appendix B Test

This test will take about an hour to complete. It includes material from all chapters. Make sure you have done all the review problems for each chapter before you take it.

1. Draw a frequency distribution for the following data:

   17, 1, 3, 4, 4, 16, 2, 2, 2, 7, 8, 7, 1, 2, 2, 5, 4, 3, 12, 9, 6, 3, 6, 10, 5, 2, 3, 11, 14, 2, 2,

2. Describe briefly the shape of the frequency distribution in problem 1.
3. What is the median of the data in problem 1?
4. What is the mode of the data in problem 1?
5. What is the mean of the data in problem 1?
6. The data in problem 1 represent a sample drawn from a larger population. What is your best estimate of the population standard deviation?
7. In a certain city 10% of the residents are over 60 years old. What is the probability of finding no one over 60 in a sample of 10 residents?
8. Establish a 95% confidence interval for μ on the basis of the following sample statistics:

   $n = 100$
   $\bar{x} = 10$
   $s = 5$

9. You are working with samples of size 15 from populations that are approximately normally distributed. What table would you use to establish confidence intervals?
10. You suspect that two brands of soft drink differ in acidity. You measure the acidity of 10 randomly selected bottles of each brand with the following results:

    Brand A   Brand B
    $n = 10$ $n = 10$
    $\bar{x} = 6.50$ $\bar{x} = 6.30$
    $s = 0.05$ $s = 0.04$

    Outline an appropriate statistical test. What assumptions must you make, if any? Is the difference between the two brands significant at the 1% level?
11. "The power of this test against the alternative $\mu = 5$ is 0.80." Explain this statement.

12. The following data represent growth rates of plants before and after a chemical treatment. Outline an appropriate statistical test of the theory that the chemical treatment reduces growth rate. What assumptions must you make, if any? Are the results significant at the 5% level?

Plant	Before	After
1	1	3
2	6	3
3	4	5
4	5	7
5	4	0
6	2	1
7	2	5

13. Compute the correlation coefficient between the before and after scores in problem 12.

14. Use a regression equation with the data of problem 12 to predict the "after" growth rate of a plant with a "before" growth rate of 3.

15. On the basis of sample data you wish to determine whether the variances of two populations differ. What statistical test will you use? Write the formula.

16. You wish to test the theory that men have higher self-esteem than women. On the basis of a questionnaire you classify a group of men and women into categories of high, medium, and low self-esteem. The results are as follows:
Perform an appropriate statistical test and comment on the results.

	High	Medium	Low
Men	10	30	10
Women	25	15	10

17. Complete the analysis of variance below. Are there significant differences between groups? ($\alpha = 0.01$)

	Sum of squares	df	Variance estimate	F
Total	580	29		
Between groups	175	2		
Within groups	405	27		

18. What assumptions are required for a two-way analysis of variance?

Answers

1.

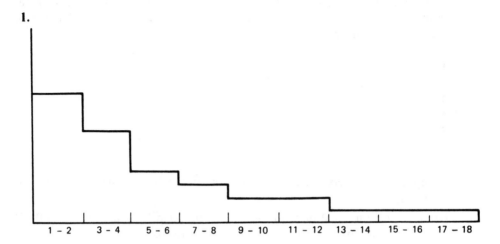

2. Skewed to the right
3. 4
4. 2
5. 5.6
6. 4.47
7. 0.349, or 35% of the time
8. μ is between 9.02 and 10.98 with 95% confidence.
9. "Critical values of t"
10. Null hypothesis $\mu_1 = \mu_2$
    Alternative $\mu_1 \neq \mu_2$
    Significance level $\alpha = 0.01$
    Critical region $t \leq -2.78$ or $t \geq +2.878$

    Assumptions: approximately normal distributions for both brands and approximately equal variances

    $t = 9.901$; the difference is significant.
11. If $\mu = 5$, you have 80% probability of obtaining significant results.
12. Use difference scores
    Null hypothesis $\mu = 0$
    Alternative $\mu < 0$
    Significance level $\alpha = 0.05$
    Critical region $t \leq -1.943$

    Assumptions: the population of difference scores is approximately normally distributed.

    $t = 0$; the results are not significant.
13. $r = 0.22$

14. $b = 0.29$; the predicted "after" score is 3.31.

15. $F = s_1^2/s_2^2$

16. Use a X^2 test. $X^2 = 11.34$, which is significant at the 1% level. This result indicates that the distribution obtained was unlikely to have occurred by chance if there were no difference between men and women, but the results do not support the theory that men have higher self-esteem than women.

17.

	Variance estimate	F
Total		
Between groups	87.5	5.83
Within groups	15.0	

The differences between groups are significant.

18. Assumptions: all observations are independent; all groups have approximately normal distributions and approximately equal variances.

References

Question	Chapter	Question	Chapter	Question	Chapter
1	1	7	2	13	7
2	1	8	3	14	7
3	1	9	3	15	6
4	1	10	5	16	8
5	1	11	4	17	6
6	3	12	5	18	9

Index